Student Study Guide

Laurel Technical Services
Richard Semmler

Elementary and Intermediate Algebra
for College Students

Allen R. Angel

Prentice
Hall

Upper Saddle River, NJ 07458

Executive Editor: Karin E. Wagner
Supplement Editor: Kate Marks
Special Projects Manager: Barbara A. Murray
Production Editor: Wendy A. Perez
Supplement Cover Manager: Paul Gourhan
Supplement Cover Designer: PM Workshop Inc.
Manufacturing Buyer: Alan Fischer

ISBN 0-13-085513-8

Prentice-Hall International (UK) Limited, London
Prentice-Hall of Australia Pty. Limited, Sydney
Prentice-Hall Canada, Inc., Toronto
Prentice-Hall Hispanoamericana, S.A., Mexico
Prentice-Hall of India Private Limited, New Delhi
Pearson Education Asia Pte. Ltd., Singapore
Prentice-Hall of Japan, Inc., Tokyo
Editora Prentice-Hall do Brazil, Ltda., Rio de Janeiro

Table of Contents

Chapter 1

1.2 Problem Solving

Summary

Guidelines for Problem Solving

1. **Understand the problem.**

    *   Read the problem *carefully* at least twice. In the first reading, get a general overview of the problem. In the second reading, determine **(a)** exactly what you are being asked to find and **(b)** what information the problem provides.

    *   Make a list of the given facts. Determine which are pertinent to solving the problem.

    *   Determine whether you can substitute smaller or simpler numbers to make the problem more understandable.

    *   If it will help you organize the information in a table.

    *   If possible, make a sketch to illustrate the problem. Label the information given.

2. **Translate the problem to mathematical language.**

    *   This will generally involve expressing the problem in terms of an algebraic expression or equation. (We will explain how to express application problems as equations in Chapter 3.)

    *   Determine whether there is a formula that can be used to solve the problem.

3. **Carry out the mathematical calculations necessary to solve the problem.**

> **4. Check the answer obtained in step 3.**
>
> - Ask yourself, "Does the answer make sense?" Is the answer reasonable?" If the answer is not reasonable, recheck your method for solving the problem and your calculations.
>
> **5. Make sure you have answered the question.**
>
> - State the answer clearly.

Example 1
Tom can pick 16 apples every minute. How many apples can be picked in an 8 hour day.

Solution
We want to find the number of apples picked in 8 hours. Since there are 60 minutes to an hour and 8 hours to the work day, the formula is

Apples picked = (16 apples per minute)(60 minutes per hour)(8 hours)
$$= (16)(60)(8)$$
$$= 7680 \text{ apples}$$

Bar, Line Graphs and Circle Graphs

Summary

> **Pie Graphs**
>
> 1. A **pie graph**, also called a **circle graph**, displays information using a circle.
>
> 2. The circle is divided into pieces called **sectors**

Example 2
Study the pie graph below. Assume that there is a total of 100,000 deaths in the 25-44 male age group.

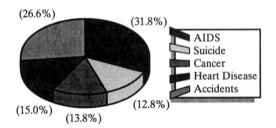

(26.6%) (31.8%)

AIDS
Suicide
Cancer
Heart Disease
Accidents

(15.0%) (13.8%) (12.8%)

    a. What should the sum of the amounts in the five sectors equal?

    b. Approximately how many deaths in this age group are caused by accidents?

    c. Approximately how many deaths in this age group are caused by cancer?

Solution

    a. Since the circle represents one whole or 100%, the sum of the five sectors should equal 1.00 (100%).

    b. Deaths caused by accident $= (26.6\%)(100,000)$
$$= (0.266)(100,000)$$
$$= 26,600$$

    c. Deaths caused by cancer $= (13.8\%)(100,000)$
$$= (0.138)(100,000)$$
$$= 13,800$$

Summary

Bar Graphs

1. A typical bar graph indicates amounts on the vertical axis.

2. The horizontal axis may indicate additional information such as time (in years or months, usually).

3. Bar graphs can be used to indicate trends.

4. The vertical axis of a bar graph should start at zero to correctly reflect the information.

Example 3
Study the bar graph which indicates the estimated construction costs of Denver's new airport.

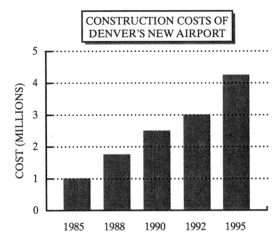

CONSTRUCTION COSTS OF
DENVER'S NEW AIRPORT

COST (MILLIONS)

1985   1988   1990   1992   1995

  a. What were the construction costs of the airport in 1985?

  b. How much more were the construction costs in 1995 than in 1985?

  c. What is the percentage change in construction costs from 1985 to 1995?

Solution

  a. Find the bar for the year 1985. This is the first bar in the graph. Look at the top of the bar and notice it stops at 1. Thus, the construction costs were 1 million dollars.

  b. First, find the estimated construction costs in 1995. Locate the year 1995 on the horizontal axis. Extend the top of the bar until it intersects the vertical axis. Extend the top of the bar until it intersects the vertical axis. The point is approximately midway between 4.0 and 4.5. So, we shall call the value 4.25. Find the difference between 4.25 and 1.0, which is 3.25. Thus, estimated construction costs were 3.25 million dollars higher in 1995 than in 1985.

  c. The percentage change can be calculated as the
$$\frac{\text{amount of change}}{\text{original amount}} \times 100\%$$

The amount of change is 3.25. The original amount is 1.0. Thus, the percentage change is $\frac{3.25}{1.00} \times 100\%$, or 325%.
This is equivalent to saying that estimated construction costs of Denver's airport was 3.25 times higher in 1995 than in 1985.

Summary

Line Graphs

1. A **line graph** generally has amounts
    indicated on the vertical axis, and some
    measure of time, such as years or months,
    listed on the horizontal axis.

2. **Line graphs** can be used to predict trends
    in data.

Example 4
Study the line graph which indicates the annual costs of regulation of American companies (in 1991 dollars).

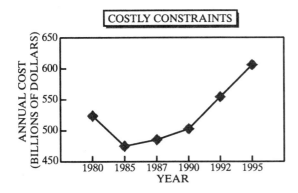

a. What type of trend does the line graph show from the years 1980 to 1985?

b. What type of trend does the line graph show from the years 1988 to 1995?

Solution

a. In 1980, the annual costs of governmental regulation was slightly more than $500 billion. In 1985, the
    annual costs of regulations dropped to slightly more than $450 billion. There was a decreasing trend in
    the costs of governmental regulation from the years 1980 to 1985.

b. From the years of 1988 to 1995, there was an increasing trend in the costs of regulations.

Summary

Statistics

1. The **mean** and **median** are two measures of central tendency.

2. The **mean** of a set of data is determined by adding all the values and dividing the sum by the number of values.

3. The **median** is the value in the middle of a set of ranked data.

Example 5
Sally is enrolled in chemistry and has taken five tests with scores of 85, 90, 82, 70, and 78

   a. Find the mean

   b. Find the median

Solution

   a. The mean is $\dfrac{\text{adding values}}{\text{sum of numbers}} = \dfrac{85+90+82+70+78}{5} = \dfrac{405}{5} = 81$

   b. The ranked values are 70, 78, 82, 85, 90 and the median is 82. The value is in the middle.

Exercise Set 1.2

   1. Study the bar graph below and answer the following questions.

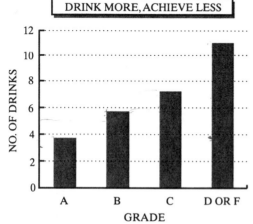

DRINK MORE, ACHIEVE LESS

   a. What is the overall trend in average number of drinks per week and grade average?

    b. What is the percentage increase in the average number of drinks per week for those students whose grade average is D or F versus those students whose grade average is A?

    c. By how much does the average of drinks per week for C students exceed the average number of drinks per week for B students?

2. Study the line graph below and answer the following questions.

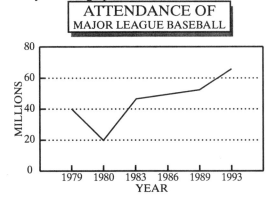

    a. What is the overall trend in attendance of major league baseball games since 1980?

    b. During what two years was the rate of growth in attendance the fastest?

3. Consider the bar graph depicting the average major league baseball team operating costs.

    a. Approximately what was the change in operating costs from 1977 to 1993?

    b. What is the overall trend in operating costs?

    c. What is the percentage increase in operating costs from 1983 to 1989?

4. Find the mean and median for the scores 17, 25, 6, 4, 31, 19, 20.

Answers to Exercise Set 1.2

  1. **a.**   As the average number of drinks per week increases, the grade average decreases.

      b.   about 200%

      c.   about 2 drinks

  2. **a.**   overall increasing trend. Every year since 1980 the attendance has increased.

      b.   from 1980–1983

  3. **a.**   Approximately $60 million.

      b.   operating costs have risen steadily from 1977 to 1993.

      c.   about 100%

  4.   mean is $\dfrac{122}{7} = 17\dfrac{3}{7}$. median is 19.

1.3 Fractions

Summary

---

  1. If a and b represent mathematical quantities, the following may represent the product of a a and b: ab, $a \cdot b$, $a(b)$, $(a)b$, $(a)(b)$.

  2. The numbers or variables multiplied in a multiplication problem are called **factors**.

  3. The top number of a fraction is called the **numerator** and the bottom number is called the **denominator**.

  4. A fraction is simplified when the numerator and denominator have no factors other than one.

Example 1
In $(7)(5) = 35$, 7 and 5 are called factors of the product 35.

Example 2
In $9x$, 9 and x are factors of the product $9x$.

Summary

> **To Simplify a Fraction**
>
> 1. Find the largest number that will divide (without remainder) both the numerator and the denominator. This number is called the **greatest common factor**.
>
> 2. Then divide both the numerator and the denominator by the greatest common factor.

Example 3

Simplify $\dfrac{24}{40}$.

Solution

The common factors of 24 and 40 are 2, 4, and 8. The greatest common factor is 8. Divide both numerator and denominator by 8 to give $\dfrac{24 \div 8}{40 \div 8} = \dfrac{3}{5}$.

The answer is $\dfrac{3}{5}$.

Example 4

Simplify $\dfrac{13}{91}$.

Solution

Since 13 evenly divides 91, we will divide both numerator and denominator by 13:

$$\dfrac{13 \div 13}{91 \div 13} = \dfrac{1}{7}$$

Summary

> **Multiplication of Fractions**
>
> Multiply the numerators and place the product over the product of the denominators.

Example 5

Multiply $\dfrac{6}{11} \cdot \dfrac{5}{12} = \dfrac{30}{132} = \dfrac{5}{22}$.

Solution

To simplify this fraction, both the numerator and denominator were divided by 6. We could have divided out this common factor of 6 first before multiplying.

Example 6

Multiply $\dfrac{19}{30} \cdot \dfrac{10}{11} = \dfrac{19}{33}$.

Solution

By dividing out the common factor of 10 first, we now have $\dfrac{19}{3} \cdot \dfrac{1}{11} = \dfrac{19}{33}$.

Summary

Division of Fractions

To divide one fraction by another, invert the divisor and proceed as in multiplication.

Division of fractions can be expressed as follows: $\dfrac{a}{b} \div \dfrac{c}{d} = \dfrac{a}{b} \cdot \dfrac{d}{c} = \dfrac{ad}{bc}$.

Example 7

Divide $\dfrac{5}{9} \div \dfrac{7}{18}$

Solution

$\dfrac{5}{9} \div \dfrac{7}{18} = \dfrac{5}{9} \cdot \dfrac{18}{7} = \dfrac{5}{1} \cdot \dfrac{2}{7} = \dfrac{10}{7}$ (Notice that we divided out the common factor of 9 before multiplying.)

Example 8

Divide $\dfrac{7}{10} \div 14$

Solution

$\dfrac{7}{10} \div 14 = \dfrac{7}{10} \cdot \dfrac{1}{14} = \dfrac{1}{10} \cdot \dfrac{1}{2} = \dfrac{1}{20}$ (Once again, notice the common factor of 7 was divided out.)

Summary

Addition and Subtraction of Fractions

To add or subtract two fractions with the same denominator, add or subtract the numerators and put the result over the common denominator.

In symbols, $\dfrac{a}{c} + \dfrac{b}{c} = \dfrac{a+b}{c}$; $\dfrac{a}{c} - \dfrac{b}{c} = \dfrac{a-b}{c}$

Example 9

Add $\dfrac{1}{12} + \dfrac{7}{12}$

Solution

$\dfrac{1}{12} + \dfrac{7}{12} = \dfrac{8}{12} = \dfrac{2}{3}.$

To simplify $\dfrac{8}{12}$, the numerator and denominator each was divided by 4.

Example 10

Subtract $\dfrac{8}{15} - \dfrac{2}{15}$

Solution

$\dfrac{8}{15} - \dfrac{2}{15} = \dfrac{6}{15} = \dfrac{6 \div 3}{15 \div 3} = \dfrac{2}{5}$

Summary

> To add (or subtract) fractions with different denominators, rewrite each fraction with the same, or common, denominator.

Example 11

Add $\dfrac{3}{4} + \dfrac{1}{9}$

Solution

$\dfrac{3}{4} + \dfrac{1}{9} = \dfrac{3}{4} \cdot \dfrac{9}{9} + \dfrac{1}{9} \cdot \dfrac{4}{4} = \dfrac{27}{36} + \dfrac{4}{36} = \dfrac{31}{36}$

Common error

When adding $\dfrac{3}{4} + \dfrac{1}{3}$, students often want to "cancel" the three's. An incorrect answer of $\dfrac{1}{4}$ would be obtained.

Dividing out a common factor in the numerator of one fraction and the denominator of another fraction can only be performed when multiplying fractions. See example below:

$\dfrac{3}{4} \cdot \dfrac{1}{3} = \dfrac{1}{4}$ (Divide out common factor of 3)

The following example illustrates the procedure of **changing a mixed number to a fraction**.

Example 12

Change $7\dfrac{3}{4}$ to a fraction

Solution
Multiply the denominator, 4, by 7 to obtain 28. To this product, add the numerator of 3. The sum of 31 represents the numerator of the fraction. The denominator of the fraction we are seeking is the same the denominator of the fraction in the mixed number, 4.

So our final answer is $\dfrac{31}{4}$.

The next example illustrates the procedure of **changing a fraction greater than 1 to a mixed number.**

Example 13

Change $\dfrac{15}{8}$ to a mixed number.

Solution

Divide 8 into 15: $\begin{array}{r} 1 \\ 8\overline{)15} \\ \underline{8} \\ 7 \end{array}$

The quotient, 1, is the whole number part of our mixed number. The remainder, 7, is the numerator of the fraction in the mixed number. The denominator in the fraction of the mixed number will be the same as the denominator of the original fraction. Thus our answer is $1\dfrac{7}{8}$.

The following examples illustrate operations with mixed numbers.

Example 14

Add $3\dfrac{3}{4} + \dfrac{1}{2}$

Solution
$$
\begin{aligned}
3\frac{3}{4} + 1\frac{1}{2} &= \frac{15}{4} + \frac{3}{2} \\
&= \frac{15}{4} + \frac{6}{4} \\
&= \frac{21}{4} \\
&= 5\frac{1}{4}
\end{aligned}
$$

Example 15

Divide $2\dfrac{1}{3} \div \dfrac{5}{6}$

Solution

$$
\begin{aligned}
2\frac{1}{3} \div \frac{5}{6} &= \frac{7}{3} \div \frac{5}{6} \\
&= \frac{7}{3} \cdot \frac{6}{5} \\
&= \frac{7}{1} \cdot \frac{2}{5} \text{ (since } 6 \div 3 = 2) \\
&= \frac{14}{5} \\
&= 2\frac{4}{5}
\end{aligned}
$$

Exercise Set 1.3

Simplify each fraction

1. $\dfrac{24}{50}$

2. $\dfrac{91}{13}$

Simplify each fraction

3. $\dfrac{13}{52}$ 4. $\dfrac{70}{35}$ 5. $\dfrac{1}{2} \cdot \dfrac{3}{4}$

6. $\dfrac{12}{5} \div \dfrac{3}{7}$ 7. $3\dfrac{1}{4} \div \dfrac{5}{6}$ 8. $\dfrac{18}{36} - \dfrac{5}{36}$

9. $\dfrac{1}{3} + \dfrac{2}{5}$ 10. $\dfrac{1}{9} - \dfrac{1}{18}$ 11. $\dfrac{5}{6} - \dfrac{2}{7}$

12. $\dfrac{4}{5} - \dfrac{2}{7}$ 13. $4\dfrac{1}{3} - 3\dfrac{1}{5}$ 14. $1\dfrac{1}{3} + \dfrac{3}{5}$

15. A board which is $28\dfrac{3}{4}$ feet long is cut into 4 equal lengths. Ignoring the width of each cut, how long is each piece?

16. A recipe for rice calls for $2\dfrac{1}{4}$ cups of rice. If you wish to make $2\dfrac{1}{2}$ times as much rice, how many cups of rice do you need?

Answers to Exercise Set 1.3

1. $\dfrac{12}{25}$

2. 7

3. $\dfrac{1}{4}$

4. 2

5. $\dfrac{3}{8}$

6. $\dfrac{28}{5} = 5\dfrac{3}{5}$

7. $\dfrac{39}{10} = 3\dfrac{9}{10}$

8. $\dfrac{13}{36}$

9. $\dfrac{11}{15}$

10. $\dfrac{1}{18}$

11. $\dfrac{23}{42}$

12. $\dfrac{18}{35}$

13. $\dfrac{17}{15} = 1\dfrac{2}{15}$

14. $\dfrac{29}{15} = 1\dfrac{14}{15}$

15. 7 ft 2 inches or $7\dfrac{3}{16}$ feet

16. $5\dfrac{5}{8}$ cups

1.4   The Real Number System

Summary

The Real Number System

1. **A set** is a collection of **elements** listed within braces.

2. **Natural numbers or counting numbers** include {1, 2, 3, 4, ...}.

3. **Whole numbers** include {0, 1, 2, 3, ...}

4. The **positive integers** or **counting numbers** are the same as the **natural numbers.**

5. **Integers** consist of numbers in the set {...–3, –2, –1, 0, 1, 2, 3 ...}

6. **Rational numbers** are numbers which can be expressed as the quotient of two integers, where the denominator of the quotient is not equal to zero.

7. **Irrational numbers** include any numbers that can be represented on a number line which are **not** rational.

8. **Real numbers** are either irrational or rational. All real numbers can be represented on the real number line.

Reminder

1. All natural numbers are whole number.

2. All natural numbers are positive integers.

3. All integers are also rational numbers.

4. All rational numbers are real.

5. No irrational number is rational.

6. All irrational numbers are real.

Example 1

Consider the set $\{-\sqrt{3},\ -1,\ -0.75,\ 3,\ 0,\ \frac{2}{3},\ -3.4,\ \sqrt{10}\}$

List elements of the set that are:

a. natural numbers   **b.** integers   **c.** irrational

d. whole   **e.** rational   **f.** real

Solution

a. natural numbers: 3   **b.** integers: –1, 3, 0   **c.** irrational: $-\sqrt{3},\ \sqrt{10}$

d. whole: 0, 3   **e.** rational: $-1, -0.75, 3, 0, \frac{2}{3}, -3.4$

f. real: $-\sqrt{3}, -1, -0.75, 3, 0, \frac{2}{3}, -3.4, \sqrt{10}$

Exercise Set 1.4

List elements of each of the following sets.

1. Natural numbers greater than 7

2. Integers between –5 and 1

3. Whole numbers between 3 and 4

4. Integers between –4 and 5

5. Positive integers less than 7

True or False:

6. $\frac{1}{3}$ is an integer.   **7.** –2 is a rational number.   **8.** 0.75 is a rational number.

9. $\sqrt{9}$ is a irrational number   **10.** $\sqrt{7}$ is real   **11.** All integers are real numbers.

12. No whole number is rational.   **13.** Some real numbers are not rational.

14. Consider the set $\left\{-\frac{2}{3},\ 0,\ -3,\ 4,\ 7\frac{1}{2},\ \sqrt{5},\ -\sqrt{11},\ 1.29,\ 308\right\}$

List those numbers that are:

    a. positive integers   **b.** whole numbers   **c.** integers

    d. rational numbers   **e.** real numbers

Answers to Exercise Set 1.4

    1. $\{8, 9, 10, 11, ...\}$ **2.** $\{-4, -3, -2, -1, 0\}$ **3.** $\varnothing$

    4. $\{-3, -2, -1, 0, 1, 2, 3, 4\}$ **5.** $\{1, 2, 3, 4, 5, 6\}$ **6.** False

    7. True   **8.** True   **9.** False

    10. True   **11.** True   **12.** False

    13. True

    14. a. $\{4, 308\}$ **b.** $\{0, 4, 308\}$ **c.** $\{0, -3, 4, 308\}$

        d. $\left\{-\dfrac{2}{3},\ 0,\ -3,\ -4,\ 7\dfrac{1}{2},\ 1.29,\ 308\right\}$

        e. $\left\{-\dfrac{2}{3},\ 0,\ -3,\ 4,\ 7\dfrac{1}{2},\ \sqrt{5},\ -\sqrt{11},\ 1.29,\ 308\right\}$

1.5   Inequalities

Summary

> **1.** The number line is used to explain inequalities.
>
> **2.** The number to the right on a number line is the greater number.
>
> **3.** The symbol for "greater than" is >.
>
> **4.** The symbol for "less than" is <.
>
> **5.** The point of the inequality symbol always "points" to the smaller number.

Example 1

Insert a > or < symbol between numbers to make a true statement.

    a. $-5 \quad 7$ $-5 < 7$, since -5 is to the left of 7

    b. $\dfrac{1}{3} \quad \dfrac{1}{4}$ $\dfrac{1}{3} > \dfrac{1}{4}$, since $\dfrac{1}{3}$ is to the right of $\dfrac{1}{4}$

c. -55 -62 $-55 > -62$, since -55 is to the right of -62

d. $-2\dfrac{1}{2}$ -2.75 $-2\dfrac{1}{2} > -2.75$, since $-2\dfrac{1}{2}$ is to the right -2.75

Summary

┌───┐

Inequalities

1. The absolute value of a number is the distance between the number and zero on a real number line.

2. The absolute value of every number is either positive or zero.

3. The negative of the absolute value of a nonzero number will always be a negative number.

└───┘

Example 2

Insert $>$, $<$, or $=$ between the two numbers to make a true statement.

<u>Answers</u>

a. $|4|$ 4 $|4| = 4$, since $|4| = 4$

b. -2 $|-3|$ $-2 < |-3|$, since $|-3| = 3$

c. -3 $-|-3|$ $-3 = -|-3|$, since $-|-3| = -(3) = -3$

d. $|-6|$ 0 $|-6| > 0$, since $|-6| = 6$ is to the right of 0

e. $-|-18|$ $-|-17|$ $-|-18| < -|-17|$, since $-18 = -|-18|$ is to the left of $-17 = -|-17|$

Exercise Set 1.5

Insert a $>$ or $<$ between the numbers to make a true statement:

1. -3.4 -3

2. $-\dfrac{2}{3}$ -1

3. $\dfrac{7}{4}$ $\dfrac{4}{7}$

4. $|-1.9|$ $|-2.3|$

5. $-|2.3|$ $-(-4.9)$

6. $\left|-\dfrac{7}{2}\right|$ $\left|-\dfrac{2}{7}\right|$

7. $\dfrac{3}{4} \cdot \dfrac{3}{4}$ $\dfrac{3}{4} + \dfrac{3}{4}$

8. $3\dfrac{1}{3} \cdot 2$ $3\dfrac{1}{3} + 2$

9. $\dfrac{7}{8} - \dfrac{1}{2}$ $\dfrac{7}{8} + \dfrac{1}{2}$

10. $-|-34|$ $-|-33|$ **11.** -0.007 -0.008 **12.** 12 $\left|\dfrac{-25}{2}\right|$

Answers to Exercise Set 1.5

1. <	**2.** >	**3.** >
4. <	**5.** <	**6.** >
7. <	**8.** >	**9.** <
10. <	**11.** >	**12.** <

1.6   Addition of Real Numbers

Summary

> 1.   You can use a number line to add signed numbers.
>
> 2.   Any 2 numbers whose sum is zero are said to be opposites of each other.
>
> 3.   **To add real numbers with the same sign,** add their absolute values. The sum has the same sign as the numbers being added.

Example 1
Add $-5 + 8$ using a number line.

Solution

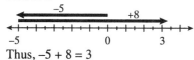

Thus, $-5 + 8 = 3$

Example 2
Add $(-7) + (-4)$ using a number line.

Solution

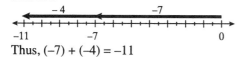

Thus, $(-7) + (-4) = -11$

Example 3
Add −6 + 6 = 0, using a number line.

Solution

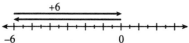

−6   0

Thus −6 + 6 = 0, since −6 + 6 = 0, we say that −6 and 6 are *opposites* of each other.

Example 4
Find the opposite of each of the following numbers:

<u>Answers</u>

   a. −3   3, since −3 + 3 = 0

   b. 4   −4, since −4 + 4 = 0

Example 5
Add 5 + 9

Solution
Since both 5 and 9 have the same sign, we add their absolute values.
$|5| + |9| = 5 + 9 = 14$

Example 6
Add −7 + (−14)

Solution
Since both numbers have the same sign, we add $|-7| + |-14| = 7 + 14 = 21$
Since both numbers being added are negative, their sum is negative so −7 + (−14) = −21.

Example 7
Add 12 + (−7)

Solution

1. Find the difference between larger absolute value and smaller
    $|12| - |-7| = 5$

2. The answer is 5 since the sign of the answer is the same as the sign of the number with the larger absolute
    value.

Example 8
Add –13 + 6

Solution

1. Find the difference between |–13| and |6| to get 13 – 6 = 7

2. The answer is –7. The sign of the answer is negative since the sign of the number with the larger absolute value is negative.

Remember:
The sum of two signed numbers with different signs may be either positive or negative. The sign of the sum will be the same as the sign of the number with the larger absolute value.

Exercise Set 1.6

Add as indicated:

1. 10 + 5	**2.** 10 + (–7)	**3.** –7 + 12
4. –12 + 5	**5.** 50 + (–35)	**6.** (–34) + (–13)
7. –103 + 73	**8.** –253 + (–147)	**9.** 59 + (–83)
10. 0 + (–43)	**11.** 12 + (–97)	**12.** –23 + (–59)
13. –138 + 6 00	**14.** –0 + 85	**15.** –0 + (–0)
16. –15 + 15	**17.** –19 + 19	**18.** –56 + 56

19. A football player gained 18 yards on one play but lost 12 yards on the second play. What is the total net yardage gained or lost?

20. A total of $622 of checks were written on a checking account. Then $1000 was deposited. What is the net change in the bank balance?

Answers to Exercise Set 1.6

1. 15	**2.** 3	**3.** 5
4. —7	**5.** 15	**6.** –47
7. –30	**8.** –400	**9.** –24
10. –43	**11.** –85	**12.** –82
13. 462	**14.** 85	**15.** 0
16. 0	**17.** 0	**18.** 0
19. 6	**20.** $378	

1.7   Subtraction of Real Numbers

Summary

If a and b represent any two real numbers, then $a - b$ = $a + (-b)$

In other words, to subtract b from a, add the **opposite** of b to a

---

In Examples 1–7, perform the additions or subtractions

Example 1
$8 - 3 = 8 + (-3) = 5$

Example 2
$5 - 11 = 5 + (-11) = -6$
(Think: $|-11| - |5| = 11 - 5 = 6$, answer is -6)

Example 3
$-9 - 3 = -9 + (-3) = -12$

Example 4
$-7 - 6 = -7 + (-6) = -13$

Example 5
Mark's checkbook balance was $235. He then wrote a check for $380, what is the new checkbook balance?
$235 - 308 = 235 + (-380) = \$-145$

In evaluating expressions involving more than one addition or subtraction, proceed from left to right unless grouping symbols are present.

Example 6
$$-6 - 8 + 15 = -6 + (-8) + 15$$
$$= -14 + 15$$
$$= 1$$

Example 7
$$-9 - (-7) + 8 - (17) = -9 + (7) + 8 + (-17)$$
$$= -2 + 8 + (-17)$$
$$= 6 + (-17)$$
$$= -11$$

Exercise Set 1.7

Evaluate the following.

  1. $7 - 4$ **2.** $-6 - 10$ **3.** $5 - 5$

4. −5 − (−9)

5. 8 − 17

6. 9 − (−14)

7. −20 − (−11)

8. 20 − (−9)

9. −20 − 9

10. (−7) − (−12)

11. −25 − 17

12. 25 − 48

13. 17 − 17

14. Subtract 8 from 19

15. Subtract −2 from −15

16. −8 + 10 − 9

17. 17 + (−13) − 10 + 18

18. 19 − 20 − 18 − 17

19. The temperature in Detroit, Michigan, fell from 32°F to −23°F. How much did the temperature drop?

20. An airplane is 2300 ft above sea level while a submarine is 1250 ft below sea level. How far above the submarine is the airplane?

Answers to Exercise Set 1.7

Evaluate the following:

1. 3

2. −16

3. 0

4. 4

5. −9

6. 23

7. −9

8. 29

9. −29

10. 5

11. −42

12. −23

13. 0

14. 11

15. −13

16. −7

17. 12

18. −36

19. 55°

20. 3550 feet

1.8 Multiplication and Division of Real Numbers

Summary

> **1.** The product of two numbers with like signs is a **positive number.**
>
> **2.** The product of two numbers with **unlike signs** is a **negative number.**
>
> **3.** Zero times any number equals 0.
>
> **4.** The product of an **even** number of negative numbers will always be **positive.**
>
> **5.** The product of an odd number of negative numbers will always be **negative.**

In Examples 1–13, perform the multipications or divisions.

Example 1
$(-5)(-4) = +20$ or 20 (like signs, product positive)

Example 2
$(-8)(14) = -112$ (unlike signs, product is negative)

Example 3
$(12)(15) = 180$ (like signs, product is positive)

Example 4
$0 \cdot 56 = 0$ (zero multiplied by **any** number is zero)

Example 5
$$\left(\frac{-5}{9}\right)\left(\frac{3}{11}\right) = \frac{(-5)(3)}{(9)(11)} = \frac{-5}{3 \cdot 11} = -\frac{5}{35}$$

Example 6
$(-5)(8)(-3)(4) = 480$

Note: The product of an even number of negative numbers will always be positive.

Example 7
$(-3)(-2)(-5)(-8)(-1)$
Think $(3)(2)(5)(8)(1) = 240$
Answer is -240 since the product of an **odd** number of negative numbers will always be negative.
Thus, $(-3)(-2)(-5)(-8)(-1) = -240$

Summary

> 1. The quotient of two real numbers with **like signs** is a **positive number.**
>
> 2. The quotient of two real numbers with **unlike signs** is a **negative number.**

Example 8
$$-35 \div 5 = \frac{-35}{5} = -7$$
Unlike signs; answer is negative.

Example 9
$$\frac{-45}{-9} = 5$$
Since -45 an d-9 have the same sign, the quotient is positive.

Example 10

$$\frac{-3}{4} \div \frac{-5}{8} = \frac{-3}{4} \cdot \frac{8}{-5} = \frac{-6}{-5} = \frac{6}{5}$$

Summary

1. For all a, b and $b \neq 0$, $\dfrac{a}{-b} = \dfrac{-a}{b} = -\dfrac{a}{b}$.

Note that $\dfrac{8}{-5} = \dfrac{-8}{5} = -\dfrac{8}{5}$

Example 11

$$\frac{4}{9} \div \frac{-5}{27} = \frac{4}{9} \cdot \frac{27}{-5}$$
$$= \frac{4(27)}{9(-5)}$$
$$= \frac{12}{-5}$$
$$= \frac{-12}{5}$$
$$= -\frac{12}{5}$$

It is customary to write the negative symbol either in the numerator or in front of the fraction.

Summary

1. $\dfrac{0}{a} = 0, \ a \neq 0$
2. $\dfrac{a}{0}$ is undefined, $a \neq 0$

Example 12

$\dfrac{0}{-17} = 0$, since $\dfrac{0}{a} = 0$ for all $a \neq 0$.

Example 13

$\dfrac{-17}{0}$ is undefined since $\dfrac{a}{0}$ is undefined for all $a \neq 0$.

Exercise Set 1.8

Find the product:

1. $8(-3)$

2. $(-17)(-9)$

3. $-1(8)$

4. $2(-4)(-9)$

5. $(-1)(-1)(-1)(17)$

6. $(-1)(50)(0)$

7. $(5)(-3)(-4)(-2)$

8. $\left(\dfrac{-5}{7}\right)\left(\dfrac{-7}{9}\right)$

9. $\left(\dfrac{-3}{4}\right)\left(\dfrac{2}{3}\right)$

10. $\left(\dfrac{-5}{7}\right)\left(\dfrac{-6}{8}\right)$

11. $\left(\dfrac{-9}{10}\right)\left(\dfrac{0}{5}\right)$

Find the quotient:

12. $\dfrac{-18}{3}$

13. $-25 \div (-5)$

14. $12 \div (-4)$

15. $\dfrac{-5}{5}$

16. $\dfrac{-91}{-13}$

17. $\dfrac{-39}{52}$

18. $\dfrac{0}{50}$

19. $\dfrac{-20}{0}$

20. $-\dfrac{16}{3} \div \dfrac{-5}{9}$

21. $-\dfrac{14}{27} \div 1$

22. $0 \div 7$

23. $7 \div 0$

Answers to Exercise Set 1.8

1. -24

2. 153

3. -8

4. 72

5. -17

6. 0

7. -120

8. $\dfrac{5}{9}$

9. $-\dfrac{1}{2}$

10. $\dfrac{15}{28}$

11. 0

12. -6

13. 5

14. -3

15. -1

16. 7

17. $-\dfrac{3}{4}$

18. 0

19. undefined

20. $\dfrac{48}{5} = 9\dfrac{3}{5}$

21. $-\dfrac{14}{27}$

22. 0

23. undefined

1.9   Exponents, Parentheses, and the Order of Operations

Summary

---

1. $a \cdot a = a^2$, $a \cdot a \cdot a = a^3$ and in general

$a \cdot a \cdot a \cdot ... \cdot a = a^n$ *n* factors of a

2. *a* is called the **base** and *n* is called **exponent**.

Example 1
Evaluate each expression

a. $4^2 = 4 \cdot 4 = 16$

b. $(-3)^3 = (-3)(-3)(-3)$
$$= (9)(-3)$$
$$= -27$$

c. $\left(\dfrac{3}{4}\right)^2 = \dfrac{3}{4} \cdot \dfrac{3}{4} = \dfrac{3 \cdot 3}{4 \cdot 4} = \dfrac{9}{16}$

d. $1^4 = 1 \cdot 1 \cdot 1 \cdot 1 = 1$

You can write algebraic expressions using exponential notation:

Example 2

a. $x \cdot x \cdot x = x^3$

b. $a \cdot a \cdot b \cdot b \cdot b = a^2 \cdot b^3$ or $a^2 b^3$

c. $3 \cdot 3 \cdot 3 \cdot (ab)(ab) = 3^3 \cdot (ab)^2$ or $3^3 (ab)^2$

d. $w \cdot w \cdot y = w^2 y^1 = w^2 y$

Whenever we see a variable raised to an exponent of 1, we can write the variable without the exponent for simplicity.

Example 3
$x^1 y^1 = xy$

Summary

To evaluate mathematical expression, the
   following order is used:

1. Evaluate expressions within **parentheses**
   first.

2. Next, evaluate all **exponential** expressions.

3. Next, evaluate all **multiplications** and
   divisions in order, working from left to
   right.

4. Finally, evaluate all **additions** and
   subtractions in the order in which they
   occur, working from left to right.

Example 6

Simplify $3 + 4^2 \cdot 5 - 7$

Solution

$$
\begin{aligned}
3 + 4^2 \cdot 5 - 7 &= 3 + 16 \cdot 5 - 7 \quad &&\text{Exponents first} \\
&= 3 + 80 - 7 \quad &&\text{Multiplication next} \\
&= 83 - 7 \\
&= 76
\end{aligned}
$$

Example 7

Simplify $(3 + 4^2) \cdot 5 - 7$

Solution

$$
\begin{aligned}
(3 + 4^2) \cdot 5 - 7 &= (3 + 16) \cdot 5 - 7 \quad &&\text{Exponents within parentheses first} \\
&= 19 \cdot 5 - 7 \quad &&\text{Work within parentheses} \\
&= 95 - 7 \quad &&\text{Multiplication before subtraction} \\
&= 88
\end{aligned}
$$

Example 8

Simplify $(9 \div 3) + 5(2 - 4)^2$

Solution

$$
\begin{aligned}
(9 \div 3) + 5(2 - 4)^2 &= (3) + 5(-2)^2 \quad &&\text{Parentheses first} \\
&= 3 + 5(+4) \quad &&\text{Exponents next} \\
&= 3 + 20 \quad &&\text{Multiplication before addition} \\
&= 23
\end{aligned}
$$

Example 9

Simplify $-5^2 + 8 \div 4$

Solution

$$
\begin{aligned}
-5^2 + 8 \div 4 &= -(5)^2 + 8 \div 4 \quad \text{Rewrite, remember that 5 is the base, not negative 3} \\
&= -25 + 8 \div 4 \quad \text{Exponents next} \\
&= -25 + 2 \quad\quad\quad \text{Multiplication before addition} \\
&= -23
\end{aligned}
$$

Example 10

Simplify $(-5)^2 + 8 \div 4$

Solution

$$
\begin{aligned}
(-5)^2 + 8 \div 4 &= 25 + 8 \div 4 \quad \text{Exponents first} \\
&= 25 + 2 \quad\quad \text{Division next} \\
&= 27
\end{aligned}
$$

Example 11

Simplify $-9 - 18 \div 9 \cdot 3^2 + 6$

Solution

$$
\begin{aligned}
-9 - 18 \div 8 \cdot 3^2 + 6 &= -9 - 18 \div 9 \cdot 9 + 6 \quad\quad \text{Exponent} \\
&= -9 - (18 \div 9) \cdot 9 + 6 \quad \text{Division and multiplication from left to right} \\
&= -9 - (2) \cdot 9 + 6 \\
&= -9 - 18 + 6 \\
&= -27 + 6 \\
&= -21
\end{aligned}
$$

Example 12

Simplify $\dfrac{5}{9} - \dfrac{3}{4} \cdot \dfrac{4}{7}$

Solution

$$
\begin{aligned}
\frac{5}{9} - \frac{3}{4} \cdot \frac{4}{7} &= \frac{5}{9} - \left(\frac{3}{4} \cdot \frac{4}{7} \right) \text{Multiplication first} \\
&= \frac{5}{9} - \frac{3}{7} \quad\quad \text{LCD is } 9 \cdot 7 = 63 \\
&= \frac{5}{9} \cdot \frac{7}{7} - \frac{3}{7} \cdot \frac{9}{9} \text{ Rewrite fractions using LCD} \\
&= \frac{35}{63} - \frac{27}{63} \\
&= \frac{8}{63}
\end{aligned}
$$

Example 13

Evaluate $8y - 17$ when $y = -2$

Solution

Substitute -2 for y

$$\begin{aligned} 8y - 17 &= 8(-2) - 17 \\ &= -16 - 17 \\ &= -33 \end{aligned}$$

Example 14

Evaluate $-x^2 + 2(y - 1) + 7$, when $x = -3$ and $y = 2$

Solution

Substitute values for variables

$$\begin{aligned} -x^2 + 2(y - 1) + 7 &= -(-3)^2 + 2(2 - 1) + 7 \\ &= -(-3)^2 + 2(1) + 7 \\ &= -(9) + 2(1) + 7 \\ &= -9 + 2 + 7 \\ &= 0 \end{aligned}$$

Exercise Set 1.9

Evaluate each expression.

1. 3^4 **2.** 1^5 **3.** $(-2)^3$

4. -7^2 **5.** -9^2 **6.** $(-9)^2$

7. $(-1)^3(4^2)$ **8.** $(-2)^3(-1)^2$ **9.** $(-2)^4(-1)^3$

10. $-2^4(-1)^2$ **11.** $4(-5)^2$

Express in exponential form:

12. $a \cdot a \cdot a \cdot a \cdot b \cdot b \cdot b$ **13.** $\left(\dfrac{1}{2}\right)\left(\dfrac{1}{2}\right)\left(\dfrac{1}{2}\right)(x \cdot x)$ **14.** $x \cdot y \cdot y \cdot y \cdot x \cdot y \cdot y \cdot x$

15. $3 \cdot aa \cdot 3 \cdot a \cdot 3b \cdot 3b$

Express as a product of factors:

16. $x^3 y^2$ **17.** $3^3 \cdot a^2 b^4$ **18.** $(-2)^2 y^4 z$

19. $(-1)^3 a^3 b^2$

Evaluate (a) $-x^2$ and (b) x^2 for each of the following values of x:

20. -3

21. 2

22. -7

23. 0

24. $-\dfrac{1}{3}$

25. $\dfrac{3}{5}$

Perform the operations and simplify

26. $5 + 2(3)$

27. $(7^2 \cdot 3) - (5 - 4)$

28. $[12 - (6 \div 3)] - 6$

29. $3^2 - 5^2(5 - 2)^2$

30. $-3^2 + 4 \cdot 9$

31. $(4^2 - 1) \div (4 + 1)^2$

32. $2.5 + 7.5 \div .3 + (.5)^2$

33. $2(5.3) + (4.3)^2 - 3.05$

34. $\dfrac{5}{8} - 5 \cdot \dfrac{1}{8}$

35. $3\left(\dfrac{4}{5} + 4\right) \div \left(\dfrac{2}{5}\right)^2$

36. $(8 + 5)^2 - (2 - 4)^2$

Write the following statement as a mathematical expression using parentheses and brackets and then evaluate:

37. Multiply 7 by 3 and to this product, add 15. Divide the sum by 8. Multiply the quotient by 9.

Evaluate the given expression for the indicated values:

38. $3x + 6$ for $x = -2$

39. $-2x^2 + 3x + 4$ for $x = -3$

40. $4(x - 3)^2$ for $x = 5$

41. $5x + 2y^2 - 6$ for $x = 2, y = -3$

42. $3x^2 - 2y^2 - 6$ for $x = 2, y = 4$

43. $-2x^2 - 4y^2 + 6(y + 1)$ for $x = -1, y = 2$

44. $4x^2 + 5x - 3$ for $x = -2$

45. $\dfrac{7x^2}{3} + \dfrac{3x^2}{2}$ for $x = 2$

46. $5x^2 + 2xy - y^2$ for $x = -3, y = 2$

Answers to Exercise Set 1.9

1. 81

2. 1

3. -8

4. -49

5. -81

6. 81

7. -16

8. -8

9. -16

10. -16

11. 100

12. $a^4 b^3$

13. $\left(\dfrac{1}{2}\right)^3 \cdot x^2$ or $\left(\dfrac{1}{2}\right)^3 x^2$ **14.** $x^3 \cdot y^5$ or $x^3 y^5$ **15.** $3^4 a^3 b^2$

16. $x \cdot x \cdot x \cdot y \cdot y$ **17.** $3 \cdot 3 \cdot 3 \cdot a \cdot a \cdot b \cdot b \cdot b \cdot b$

18. $(-2)(-2)(y)(y)(y)(y)(z)$ **19.** $(-1)(-1)(-1) \cdot a \cdot a \cdot a \cdot b \cdot b$

20. a. -9 **b.** 9

21. a. -4 **b.** 4

22. a. -49 **b.** 49

23. a. 0 **b.** 0

24. a. $-\dfrac{1}{9}$ **b.** $\dfrac{1}{9}$

25. a. $-\dfrac{9}{25}$ **b.** $\dfrac{9}{25}$

26. 11 **27.** 146 **28.** 3

29. -216 **30.** 27 **31.** $\dfrac{3}{5}$

32. 27.75 **33.** 26.04 **34.** 0

35. 90 **36.** 165 **37.** $\dfrac{81}{2}$

38. 0 **39.** -23 **40.** 16

41. 22 **42.** -26 **43.** 0

44. 3 **45.** $\dfrac{46}{3}$ **46.** 29

1.10 Properties of the Real Number System

Summary

> 1. **The commutative property of addition** states: If a and b are two real numbers then $a + b = b + a$.
>
> 2. **The commutative property of multiplication** states: If a and b are two real numbers, then $a \cdot b - b \cdot a$.
>
> 3. The commutative property **does not** hold for subtraction or division.
>
> 4. The **associative property of addition** states: If a, b and c represent any three real numbers, then $(a + b) + c = a + (b + c)$.
>
> 5. The **associative property of multiplication** states: If a, b and c are any three real numbers, then $(a \cdot b) \cdot c = a \cdot (b \cdot c)$.
>
> 6. The **distributive property** states: If a, b and c represent any three real numbers, then $a(b + c) = ab + ac$.

Hints:

1. The commutative property involve changes in order while the associative properties involve changes in grouping.

2. The distributive property involves two operations: Multiplication and addition.

For each of the examples below, name the property illustrated.

Example 1   <u>Answers</u>
$3(x + 4) = 3x + (3 \cdot 4)$ Distributive property

Example 2
$5 + (-3) = -3 + 5$ Commutative of addition

Example 3
$(5 \cdot 7) \cdot x = 5 \cdot (7 \cdot x)$ Associative property of multiplication

Example 4
$(x + 3) + 7 = (3 + x) + 7$ Commutative property of addition

Example 5
$5x + 6x = (5 + 6)x$ Distributive property (in reverse order)

Example 6
Name the property used to go from one step to the next.
                 <u>Answers</u>

$8 + 5(x + 3)$

$= 8 + 5x + 5(3)$ Distributive

$= 8 + 5x + 15$ Arithmetic fact

$= 8 + 15 + 5x$ Commutative property of addition

$= (8 + 15) + 5x$

$= 23 + 5x$ Addition fact

$= 5x + 23$ Commutative property of addition

Example 7
Name the property to go from one step to the next
                 <u>Answers</u>

$-1(3x + 2) + 2(3x + 6)$

$= -1(3x) + -1(2) + 2(3x) + 2(6)$ Distributive property

$= (-1 \cdot 3)x + -1(2) + (2 \cdot 3)x + 2 \cdot 6$ Associative property of Multiplication

$= -3x + (-2) + 6x + 12$ Arithmetic facts

$= -3x + 6x + (-2) + 12$ Commutative property of addition

$= (-3 + 6)x + (-2) + 12$ Distributive property
     $3x + 10$ Arithmetic facts

Exercise Set 1.10

Name the property illustrated:

    1. $4(2 + 7) = 4(2) + 4(7)$

    2. $x \cdot a = a \cdot x$

    3. $-1(x + 5) = -1(x) + (-1)(5)$

    4. $x + (2 + 3) = (x + 2) + 3$

    5. $(x \cdot 2) \cdot 3 = x \cdot (2 \cdot 3)$

    6. $(a + b) + x = (b + a) + x$

Name the property illustrated.

    7. $2(3x + 4y) = 2(3x) + 2(4y)$

Complete each exercise using the given property.

8. $x + 3$	**9.** $2(3x + 5)$	**10.** $(x + 3) + 9$
Commutative	Distributive	Associative

Name the property used to go from one step to the next.

    11. $8 + 3(2x + 4)$

        a. $= 8 + 3(2x) + 3(4)$

        b. $= 8 + (3 \cdot 2)\, x + 3(4)$

        c. $= 8 + 6x + 12$

        d. $= 8 + 12 + 6x$

        e. $= 20 + 6x$

        f. $= 6x + 20$

Answers to Exercise Set 1.10

    1. Distributive property

    2. Commutative property of multiplication

    3. Distributive property

    4. Associative property of addition

    5. Associative of multiplication

6. Commutative property of addition

7. Distributive property

8. $3 + x$ **9.** $2(3x) + 2(5)$ **10.** $x + (3 + 9)$

11. **a.** Distributive property

     b. Associative property of multiplication

     c. Arithmetic fact

     d. Commutative property of addition

     e. Arithmetic fact

     f. Commutative property of addition

Chapter 1 Practice Test

1. Consider the set

$$S = \left\{ -7,\ 58,\ -2\frac{1}{2},\ 0,\ 4.58,\ \sqrt{7},\ \frac{7}{8},\ -3,\ -10,\ \sqrt{9} \right\}$$

List the elements that are:

     a. natural numbers   **b.** whole numbers   **c.** integers

     d. rational numbers   **e.** irrational numbers   **f.** real numbers

2. True or false: Every integer is a rational number.

3. True or false: All whole numbers are also natural numbers.

Insert a >, <, or = in the blank to make each statement true.

4. -4 _____ -5 **5.** $-|-4|$ _____ $-|-5|$

6. -2^2 _____ 4 **7.** $|3-7|$ _____ $|7-3|$

Evaluate each expression:

8. $2 \cdot 3 + 4$ **9.** $-2 - 4 - 6$ **10.** $(-4)(3)(-2)(-2)$

11. $\left(-14 \div \frac{1}{7} \right) - (-2)$ **12.** $2 \cdot 3^2 - 4 \cdot 5^2$ **13.** $\left(\frac{-3}{4} \right)^2$

14. $-7(-2-5) \div 7$ **15.** $-9.1 \div 0.13$ **16.** $(-2)^5$

17. Write $3 \cdot 3 \cdot 5 \cdot 5 \cdot a \cdot a \cdot bbb$ in exponential form.

18. Write $3^3 \cdot 4^2 \cdot u^4 v$ as a product of factors.

Evaluate each expression for the indicated values.

19. $3x^2 - 4x$ for $x = -2$ **20.** $-x^2 + 3x + 7$ for $x = 3$

21. $3x - 2y^2 + 4$ for $x = -2, y = -2$

Name the property illustrated (22 – 25).

22. $2(3x + 2) = 2(3x) + 2(2)$ **23.** $2 + 3y = 3y + 2$ **24.** $2(3 \cdot z) = (2 \cdot 3)z$

25. $3 + (4 + y) = (3 + 4) + y$

26. Study the bar graph depicting Professor X's grade distribution for Calculus I.

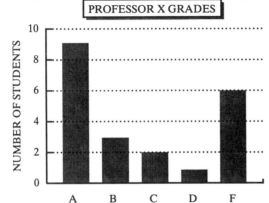

a. If all students in Professors X's class are depicted in the bar graph, how many students are enrolled in this class?

b. What percentage of the students earned an A grade?

c. What percentage of the students passed the class (pass means a grade of C or higher)

d. How many more A's than F's were earned?

Answers to Chapter 1 Practice Test

1. a. $58, \sqrt{9}$ **b.** $58, 0, \sqrt{9}$

c. $-7, 58, 0. -3, -10, \sqrt{9}$ **d.** $-7, 58, -2\frac{1}{2}, 0, 4.58, \frac{7}{8}, -3, -10, \sqrt{9}$

e. $\sqrt{7}$ **f.** $-7, 58, -2\frac{1}{2}, 0, 4.58, \sqrt{7}, \frac{7}{8}, -3, -10, \sqrt{9}$

2. True   **3.** False   **4.** $>$

5. >

6. <

7. =

8. 10

9. −12

10. −48

11. −96

12. −82

13. $\dfrac{9}{16}$

14. 7

15. −70

16. −32

17. $3^2 5^2 a^2 b^3$

18. $3 \cdot 3 \cdot 3 \cdot 4 \cdot 4 \cdot u \cdot u \cdot u \cdot u \cdot v$

19. 20

20. 7

21. −10

22. distributive property

23. commutative property of addition

24. associative property of multiplication

25. associative property of addition

26. a. 21 students

    b. 42.8%

    c. 66.67%

    d. 3

Chapter 2

2.1 Combining Like Terms

Summary

Variables are letters used to represent numbers.

An **expression** (or **algebraic expression**) is a collection of numbers, variables, grouping symbols, and operation symbols.

Terms refer to the parts of an algebraic expression that are added or subtracted.

The **coefficient** (or **numerical coefficient**) is the numerical part of a term. If a term appears without a numerical coefficient, we assume that the coefficient is 1.

A **constant** (or **constant term**) is one which contains no variables, consisting only of a number.

Like terms are terms that have the same variables with the same exponents.

Example 1

Identify any like terms in the following algebraic expressions.

<u>Answers</u>

a. $5y - x + 3y$ $\qquad$ $5y$, $3y$ are like terms

b. $4x^3 + x + x^2 + x^3$ $\qquad$ $4x^3$, x^3 are like terms

c. $ab^2 + ab + 5ab^2 - 2ab^2$ $\qquad$ ab^2, $5ab^2$, $-2ab^2$ are like terms

d. $x^2 + 2x + y^2$ $\qquad$ no like terms

Summary

> **To combine like terms:**
>
> 1. Determine which terms are like terms.
>
> 2. Add or subtract the **coefficients** of like terms.
>
> 3. Multiply the number found in step 2 by the common variables.

Example 2
Combine like terms: $5x + 4x$

Solution

    a. Determine like terms: $5x, 4x$

    b. Add coefficients of like terms: $5 + 4 = 9$

    c. Multiply the number found in **b)** by the common variable. Therefore, $5x + 4x = 9x$.

Example 3
Combine like terms: $\dfrac{3}{4}x - \dfrac{5}{6}x$

Solution

    a. Determine like terms: $\dfrac{3}{4}x, \dfrac{5}{6}x$

    b. Subtract coefficients of like terms:
$$\frac{3}{4} - \frac{5}{6} = \frac{9}{12} - \frac{10}{12} = -\frac{1}{12}$$

    c. Multiply the number found in b) by the common variable. Therefore, $\dfrac{3}{4}x - \dfrac{5}{6}x = -\dfrac{1}{12}x$.

Example 4
Combine like terms: $4y + 7 - 7y$

Solution: We can rearrange as $4y - 7y + 7$

    a. Determine like terms: $4y, 7y$

    b. Subtract coefficients of like terms: $4 - 7 = -3$

    c. Multiply the numbers found in b) by the common variable. Therefore, $4y + 7 - 7y = -3y + 7$.

Example 5
Combine like terms: $2x + 4y + 6 - 3y + 7x - 8$

Solution
Again, rearrange as $2x + 7x + 4y - 3y + 6 - 8$.

    a. Determine like terms: $2x, 7x$
$$4y, 3y$$
$$6, 8$$

    b. Add or subtract the coefficients of like terms.

        For x: $2 + 7 = 9$

        For y: $4 - 3 = 1$

    c. Multiply the numbers found step 2 by the common variables. Therefore,
$2x + 4y + 6 - 3y + 7x - 8 = 9x + y - 2$

Summary

Distributive Property

For any real numbers $a, b, c,$
$$a(b + c) = ab + ac.$$

Example 6
Use the distributive property to remove parentheses.

    a. $3(x + 5) = 3x + 3(5)$
$$= 3x + 15$$

    b. $-4(x - 2) = -4x + (-4)(-2)$
$$= -4x + 8$$

    c. $2(3x + 5) = 2(3x) + 2(5)$
$$= 6x + 10$$

Note: A common student error is to fail to distribute over the second term in parentheses.

Summary

The distributive property can be extended as follows:

$$a(b + c + d + ... + n) = ab + ac + ad + ... + an$$

Example 7
Use the distributive property to remove parentheses.

a. $2(x - y + 5) = 2x + 2(-y) + 2(5)$
$$= 2x - 2y + 10$$

b. $3(x^2 + 4x - 5) = -3(x^2) + (-3)(4x) + (-3)(-5)$
$$= -3x^2 - 12x + 15$$

Summary

When no sign or a plus sign precedes parentheses, the parentheses may be removed without having to change the expression inside the parentheses.

When a negative sign precedes parentheses, the signs of all terms within the parentheses are changed when the parentheses are removed.

Example 8
Remove parentheses.

Answers

a. $(x - 3)$ $x - 3$

b. $-(x - 3)$ $-x + 3$

c. $-(x^2 + 5)$ $-x^2 - 5$

d. $-(x^2 + 4x - 3)$ $-x^2 - 4x + 3$

Summary

To simplify an expression:

1. Use the distributive property to remove any parentheses.

2. Combine like terms.

Example 9
Simplify the following.

a. $5 - (4x + 2) = 5 - 4x - 2$ Distributive property

$$= -4x + 3$$ Combine like terms.

b. $5y + 3(y - 4) = 5y + 3y - 12$ Distributive property

$= 8y - 12$ Combine like terms.

c. $-3(x + 3) - 4 = -3x - 9 - 4$ Distributive property

$= -3x - 13$ Combine like terms.

d. $4(x + 2y) + 2(x - 3y)$

$= 4x + 8y + 2x - 6y$ Distributive property

$= 4x + 2x + 8y - 6y$ Rearrange terms.

$= 6x + 2y$ Combine like terms.

Exercise Set 2.1

Combine like terms when possible.

1. $3x - 6x$ **2.** $x + y + 2z$ **3.** $3x - 2x - 4x$

4. $-2x + 3y - 7y$ **5.** $5x - 8 + 4$ **6.** $\dfrac{2}{3}x + \dfrac{3}{4} + \dfrac{3}{4}x$

7. $0.18x + 1.8x$ **8.** $3x^2 - 8x + 5x^2 - 6x$

Use the distributive property to remove parentheses.

9. $4(x - 5)$ **10.** $-3(x + 6)$ **11.** $6(x^2 - 2x + 1)$

12. $2(x - y - 5)$ **13.** $-2(-y^2 + 3y - 2)$ **14.** $\dfrac{3}{4}(4x - 8)$

Simplify when possible.

15. $-(2x^2 - x - 5)$ **16.** $2x + 3(x - 5)$ **17.** $4a - (6a - b)$

18. $7y - 2(y - 3)$ **19.** $5 + (7 - y) + y$ **20.** $2(2x + y) - 3(x + 4y)$

Answers to Exercise Set 2.1

1. $-3x$ **2.** $x + y + 2z$ **3.** $-3x$

4. $-2x - 4y$ **5.** $5x - 4$ **6.** $\dfrac{17}{12}x + \dfrac{3}{4}$

7. $1.98x$ **8.** $8x^2 - 14x$ **9.** $4x - 20$

10. $-3x - 18$ **11.** $6x^2 - 12x + 6$ **12.** $2x - 2y - 10$

13. $2y^2 - 6y + 4$ 14. $3x - 6$ 15. $-2x^2 + x + 5$

16. $5x - 15$ 17. $-2a + b$ 18. $5y + 6$

19. 12 20. $x - 10y$

2.2 The Addition Property of Equality

Summary

> An **equation** is a statement that shows two algebraic expressions are equal.
>
> A **linear equation** in one variable is an equation which can be written in the form:
> $ax + b = c$, for real numbers, a, b, and c, $a \neq 0$.
>
> A **solution** of an equation is the number or numbers that make the equation true.

Example 1
Consider the equation $3x + 2 = 8$.

   a. Determine whether $x = 3$ is a solution.

     Substitute 3 for x

$$3(3) + 2 = 8$$
$$9 + 2 = 8$$
$$11 = 8 \quad \text{False}$$

     Therefore, 3 is not a solution.

   b. Determine whether $x = 2$ is a solution.

     Substitute 2 for x

$$3(2) + 2 = 8$$
$$6 + 2 = 8$$
$$8 = 8 \quad \text{True}$$

     Therefore, 2 is a solution to the equation.

Example 2
Consider the equation $4(x-3) - 5x = -12$.

Determine whether $x = 0$ is a solution.

Substitute 0 for x.

$$4(0-3) - 5(0) = -12$$
$$4(-3) - 5(0) = -12$$
$$-12 - 0 = -12$$
$$-12 = -12 \text{ True}$$

Therefore, $x = 0$ is a solution to the equation.

Summary

Addition Property of Equality

If $a = b$, then $a + b = a + c$ for any real number a, b, and c. Since subtraction is defined in terms of addition, the addition property also allows us to subtract the same number from both sides of the equation.

Example 3
Solve the following equations.

    a. $x - 6 = 4$ **b.** $x + 5 = 3$

    c. $x - 0.6 = 0.2$ **d.** $-3 = y + 6$

Solution

    a. $x - 6 = 4$

          $x - 6 + 6 = 4 + 6$ Add 6 to both sides

          $x + 0 \quad = 10$

          $x \qquad\quad = 10$

    b. $x + 5 = 3$

          $x + 5 + (-5) = 3 + (-5)$ Add (-5) to both sides

          $x + 0 \qquad = -2$

          $x \qquad\qquad = -2$

For the above equation, it could be done in the following manner:

$$x + 5 \quad = 3$$
$$x + 5 - 5 = 3 - 5 \qquad \text{Subtract 5 from both sides}$$
$$x + 0 \quad = -2$$
$$x \quad\quad = -2$$

c. $\quad x - 0.6 \quad\quad = 0.2$
$$x - 0.6 + 0.6 = 0.2 + 0.6 \quad \text{Add 0.6 to both sides}$$
$$x + 0 \quad\quad\quad = 0.8$$
$$x \quad\quad\quad\quad = 0.8$$

d. $\quad -3 \quad\quad = y + 6$
$$-3 + (-6) = y + 6 + (-6) \qquad \text{Add } (-6) \text{ to both sides}$$
$$-9 \quad\quad = y + 0$$
$$-9 \quad\quad = y$$

Exercise Set 2.2

Solve each equation and check your solution.

1. $5 + x = 8$ **2.** $-3 = y - 7$ **3.** $x + 8 = -2$

4. $-6 = -8 + a$ **5.** $0.2 + c = 0.8$ **6.** $-0.5 = x + 0.5$

7. $-7 = x - 7$ **8.** $2 + y = 0$ **9.** $a - 1.6 = 5.3$

10. $-2.6 = -4.3 + x$

Answers to Exercise Set 2.2

1. 3   **2.** 4   **3.** −10

4. 2   **5.** 0.6   **6.** −1

7. 0   **8.** −2   **9.** 6.9

10. 1.7

2.3   The Multiplication Property of Equality

Summary

> Two numbers are **reciprocals** of each other when their product is 1.

Example 1
Give the reciprocals of the following numbers.

<u>Answers</u>

a. 5 $\dfrac{1}{5}$ is reciprocal since $\dfrac{1}{5} \cdot 5 = 1$

b. −6 $-\dfrac{1}{6}$ is reciprocal since $-\dfrac{1}{6} \cdot -6 = 1$

c. $-\dfrac{2}{3}$ $-\dfrac{3}{2}$ is reciprocal since $-\dfrac{2}{3} \cdot -\dfrac{3}{2} = 1$

Summary

Multiplication Property of Equality

If $a = b$, then $a \cdot c = b \cdot c$ for any numbers a, b, and c.

The Multiplication Property can be used to solve equations of the form $ax = b$.

Example 2

Solve the following equations.

a. $4x = 20$ **b.** $42 = 3x$ **c.** $\dfrac{x}{3} = 5$

d. $-\dfrac{3}{4}x = 9$ **e.** $1.44 = \dfrac{x}{3}$

Solution

a. $4x = 20$

$\dfrac{1}{4} \cdot 4x = \dfrac{1}{4} \cdot 20$ Multiply both sides by $\dfrac{1}{4}$

$1 \cdot x = 5$

$x = 5$

b. $42 = 3x$

$\dfrac{1}{3} \cdot 42 = \dfrac{1}{3} \cdot 3x$ Multiply both sides by $\dfrac{1}{3}$

$14 = 1 \cdot x$

$14 = x$

47

c. $\dfrac{x}{3} = 5$

$\dfrac{3}{1} \cdot \dfrac{x}{3} = \dfrac{3}{1} \cdot 5$ Multiply both sides by $3 \left(\text{or } \dfrac{3}{1} \right)$

$x = 15$

d. $-\dfrac{3}{4}x = 9$

$\left(-\dfrac{4}{3} \right)\left(-\dfrac{3}{4}x \right) = \left(-\dfrac{4}{3} \right)(9)$ Multiply both sides by $-\dfrac{4}{3}$

$1 \cdot x = -12$

$x = -12$

e. $1.44 = \dfrac{x}{3}$

$3(1.44) = 3 \cdot \dfrac{x}{3}$ Multiply both sides by 3

$4.32 = 1 \cdot x$

$4.32 = x$

Summary

> Since division can be defined in terms of multiplication $\left(\dfrac{a}{b} \text{ means } a \cdot \dfrac{1}{b} \right)$, the multiplication property also allows us to divide both sides of an equation by the same nonzero numbers.

Example 3
Solve the following equations.

   a. $6x = -3$ **b.** $0.16x = 0.8$

Solution

a. $6x = -3$

$\dfrac{6x}{6} = -\dfrac{3}{6}$ Divide both sides by 6

$1 \cdot x = -\dfrac{3}{6}$

$x = -\dfrac{1}{2}$

b. $0.16x = 0.8$

$$\frac{0.16x}{0.16} = \frac{0.8}{0.16} \qquad \text{Divide both sides by } 0.16$$

$$1 \cdot x = 5$$

$$x = 5$$

Note: A common student error is to obtain the equation $-x = 5$ and believe that the equation is solved. **The solution must be of the form** $x =$ some number.

For any real number, a, $a \neq 0$, if $-x = a$, then $x = -a$. Therefore, if $-x = 3$, then $x = -3$. If $-x = -6$, then $x = -(-6) = 6$.

Exercise Set 2.3

Solve each equation and check your solution.

1. $3x = 15$

2. $-4x = 24$

3. $7x = 28$

4. $\dfrac{x}{3} = 5$

5. $\dfrac{x}{-4} = 6$

6. $-6 = \dfrac{x}{7}$

7. $\dfrac{3}{4}x = -12$

8. $-\dfrac{4}{5}x = 10$

9. $-7 = -\dfrac{2}{7}x$

10. $100 = -25x$

11. $-56 = -14x$

12. $0.25x = 1$

13. $-x = -(-8)$

14. $-\dfrac{1}{3}x = -6$

15. $-\dfrac{1}{4}x = \dfrac{2}{3}$

Answers to Exercise Set 2.3

1.   5

2.   –6

3.   4

4.   15

5.   –24

6.   –42

7.   –16

8. $-\dfrac{25}{2}$

9. $\dfrac{49}{2}$

10.   –4

11.   4

12.   4

13.   –8

14.   18

15. $-\dfrac{8}{3}$

2.4   Solving Linear Equations with a Variable on Only One Side of the Equation

Summary

To solve Linear Equations with a Variable on only one side of the Equation:

1. Use the distributive property to remove parentheses.

2. Combine like terms on the same side of the equal sign.

3. Use the addition property to obtain an equation with the term containing the variable on one side of the equal sign and a constant on the other side. This will result in an equation of the form $ax = b$.

4. Use the multiplication property to isolate the variable. This will give a solution of the form $x = \dfrac{b}{a}$ or $1 \cdot x = \dfrac{b}{a}$.

5. Check the solution in the original equation.

Example 1
Solve the equation $4x + 3 = 11$.

Solution

$$4x + 3 = 11$$

(step 3) $4x + 3 - 3 = 11 - 3$ Subtract 3 from both sides

$$4x = 8$$

(step 4) $\dfrac{4x}{4} = \dfrac{8}{4}$ Divide both sides by 4

$$x = 2$$

(step 5) Check: $4x + 3 = 11$

$$4(2) + 3 = 11$$

$$8 + 3 = 11$$

$$11 = 11 \text{ True}$$

Example 2
Solve the equation $-8 = -6x - 26$.

Solution

$$-8 = -6x - 26$$

(step 3) $\quad -8 + 26 = -6x - 26 + 26 \quad$ Add 26 to both sides

$$18 = -6x$$

(step 4) $\quad \dfrac{18}{-6} = \dfrac{-6x}{-6} \qquad$ Divide both sides by -6

$$-3 = x$$

(step 5)　Check: $\quad -8 = -6x - 26$

$$-8 = -6(-3) - 26$$

$$-8 = 18 - 26$$

$$-8 = -8 \text{ True}$$

Example 3
Solve the equation $7x + 3 - 2x = -22$.

Solution

$$7x + 3 - 2x = -22$$

(step 2) $\qquad 5x + 3 = -22 \qquad$ Combine like terms

(step 3) $\quad 5x + 3 - 3 = -22 - 3 \quad$ Subtract 3 from both sides

$$5x = -25$$

(step 4) $\qquad \dfrac{5x}{5} = \dfrac{-25}{5} \qquad$ Divide both sides by 5

$$x = -5$$

(step 5) Check: $\qquad 7x + 3 - 2x = -22$

$$7(-5) + 3 - 2(-5) = -22$$

$$-35 + 3 + 10 = -22$$

$$-35 + 13 = -22$$

$$-22 = -22 \text{ True}$$

Example 4
Solve the equation $x + 1.53 + 0.16x = 3.85$.

Solution

$$x + 1.53 + 0.16x = 3.85$$

(step 2) $\qquad 1.16x + 1.53 = 3.85 \qquad$ Combine like terms

(step 3) $1.16x + 1.53 - 1.53 = 3.85 - 1.53 \quad$ Subtract 1.53 from both sides

$$1.16x = 2.32$$

(step 4) $\qquad \dfrac{1.16x}{1.16} = \dfrac{2.32}{1.16} \qquad$ Divide both sides by 1.16

$$x = 2$$

51

(step 5) Check: $x + 1.53 + 0.16x = 3.85$

$2 + 1.53 + 0.16(2) = 3.85$

$2 + 1.53 + 0.32 = 3.85$

$3.85 = 3.85$ True

Example 5

Solve the equation $-3(x - 5) + 2x = 5$.

Solution

$-3(x - 5) + 2x = 5$

(step 1) $-3x + 15 + 2x = 5$ Distributive property

(step 2) $-x + 15 = 5$ Combine like terms

(step 3) $-x + 15 - 15 = 5 - 15$ Subtract 15 from both sides

$-x = -10$

$x = 10$

(Step 4) Check: $-3(x - 5) + 2x = 5$

$-3(10 - 5) + 2(10) = 5$

$-3(5) + 2(10) = 5$

$-15 + 20 = 5$

$5 = 5$ True

Exercise Set 2.4

Solve the following equations.

1. $4x - 6 = 10$

2. $3x + 7 = 10$

3. $-12 = 6x + 12$

4. $2x - 3 + 5x = 11$

5. $-2 = -2x - 6 - 2x$

6. $10x + 3 - 4x = 7 - 16$

7. $-2x + 3 = -9$

8. $-\dfrac{3}{4}x - 5 = 4$

9. $0.6 + 0.6x = -1.8$

10. $x - 2x = 6$

11. $-7 = -(x - 4)$

12. $4(x - 3) = 16$

13. $x - (4x + 5) = -11$

14. $10 = 2(x + 5) - 6x$

15. $4(x + 3) + 2(2x - 6) = 32$

16. $3(2x - 5) + 2x = -23$

17. $x - 0.35x = 1.95$

18. $0.1(2x - 3) = 0$

Answers to Exercise Set 2.4

1. 4

2. 1

3. -4

4. 2

5. -1

6. -2

7. 6

8. -12

9. -4

10. -6

11. 11

12. 7

13. 2

14. 0

15. 4

16. -1 **17.** 3 **18.** $\dfrac{3}{2}$ or 1.5

2.5   Solving Linear Equations with the Variable on Both Sides of the Equation

Summary

To Solve Linear Equations with the Variable on Both Sides of the Equal Sign:

1. Use the distributive property to remove parentheses.

2. Combine like terms on the same side of the equal sign.

3. Use the addition property to rewrite the equation with all terms containing the variable on one side of the equal sign and all terms not containing the variable on the other side of the equal sign. It may be necessary to use the addition property to accomplish this goal. You will eventually get an equation of the form $ax = b$.

4. Use the multiplication property to isolate the variable. This will give a solution of the form $x =$ some number.

5. Check the solution to the original solution.

Example 1
Solve the equation $7y - 4 = 4y + 20$.

Solution

	$7y - 4 = 4y + 20$	Steps 1 and 2 are not needed
(step 3)	$7y - 4 - 4y = 4y - 4y + 20$	Subtract $4y$ from both sides
(step 3)	$3y - 4 + 4 = 20 + 4$	Add 4 to both sides
	$3y = 24$	
(step 4)	$\dfrac{3y}{3} = \dfrac{24}{3}$	Divide both sides by 3
	$y = 8$	

$$\text{(Step 5) Check: } 7y - 4 = 4y + 20$$
$$7(8) - 4 = 4(8) + 20$$
$$56 - 4 = 32 + 20$$
$$52 = 52 \qquad \text{True}$$

Example 2
Solve the equation $4r - 5r - 15 = 2r$.

Solution

$$4r - 5r - 15 = 2r$$

(step 2) $-r - 15 = 2r$ Combine like terms

(step 3) $-r + r - 15 = r + 2r$ Add r to both sides

$$-15 = 3r$$

(step 4) $-\dfrac{15}{3} = \dfrac{3r}{3}$ Divide both sides by 3

$$-5 = r$$

(step 5) Check: $4r - 5r - 15 = 2r$

$$4(-5) - 5(-5) - 15 = 2(-5)$$

$$-20 + 25 - 15 = -10$$

$$-10 = -10 \quad \text{True}$$

Example 3
Solve the equation $3 - v + 7 = -1 - 2(2 - v)$.

Solution

$$3 - v + 7 = -1 - 2(2 - v)$$

(step 1) $3 - v + 7 = -1 - 4 + 2v$ Distributive property

(step 2) $10 - v = -5 + 2v$ Combine like terms

(step 3) $10 - 10 - v = -5 - 10 + 2v$ Subtract 10 from both sides

$$-v = -15 + 2v$$

(step 3) $-v - 2v = -15 + 2v - 2v$ Subtract $2v$ from both sides

$$-3v = -15$$

(step 4) $\dfrac{-3v}{-3} = \dfrac{-15}{-3}$ Divide both sides by -3

$$v = 5$$

(step 5) Check: $3 - v + 7 = -1 - 2(2 - v)$

$$3 - 5 + 7 = -1 - 2(2 - 5)$$

$$5 = -1 - 2(-3)$$

$$5 = 5 \qquad \text{True}$$

Example 4
Solve the equation $0.1(x - 20) = 0.4x + 1$.

Solution:

$$0.1(x - 20) = 0.4x + 1$$

(step 1) $\quad 0.1x - 2 = 0.4x + 1 \qquad$ Distributive property

(step 3) $\quad 0.1x - 0.4x - 2 = 0.4x - 0.4x + 1 \qquad$ Subtract $0.4x$ from both sides

$$-0.3x - 2 = 1$$

(step 3) $\quad -0.3x - 2 + 2 = 1 + 2 \qquad$ Add 2 to both sides

(step 4) $\quad \dfrac{-0.3x}{-0.3} = \dfrac{3}{-0.3} \qquad$ Divide both sides by -0.3

$$x = -10$$

(step 5) Check: $0.1(x - 20) = 0.4x + 1$

$$0.1(-10 - 20) = (0.4)(-10) + 1$$

$$0.1(-30) = -4 + 1$$

$$-3 = -4 + 1$$

$$-3 = -3 \qquad \text{True}$$

Summary

Equations that are true for all values of x are called **identities**.

Equations for which you obtain an obviously false statement have no solution.

Example 5
Solve the equation $3(x - 2) = 6x - 3x - 6$.

Solution

$$3(x - 2) = 6x - 3x - 6$$

(step 1) $\quad 3x - 6 = 6x - 3x - 6 \qquad$ Distributive property

(step 2) $\quad 3x - 6 = 3x - 6 \qquad$ Combine like terms

(step 3) $\quad 3x - 3x - 6 = 3x - 3x - 6 \qquad$ Subtract $3x$ from both sides

$$-6 = -6$$

(step 3) $\quad -6 + 6 = -6 + 6 \qquad$ Add 6 to both sides

$$0 = 0$$

Therefore, since $0 = 0$ is always true, the equation is true for all values of x.

Example 6
Solve the equation $2x + 19 - x = 8x + 3 - 7x$.

Solution
$$2x + 19 - x = 8x + 3 - 7x$$

(step 2) $x + 19 = x + 3$ Combine like terms

(step 3) $x - x + 19 = x - x + 3$ Subtract x from both sides

$$19 = 3$$

Since $19 = 3$ is false, the equation has no solution.

Exercise Set 2.5

Solve each equation.

1. $6y + 2 = y + 17$ 2. $11n + 3 = 10n + 11$ 3. $3x + 1 = 7$

4. $4x - 7 = 5x + 1$ 5. $5a + 7 = 2a + 7$ 6. $6a + 3 + 2a = 11$

7. $5y + 9 + 2y = 30 - 7$ 8. $5x + 2(x + 1) = 23$ 9. $6y + 2(2y + 3) = 16$

10. $\frac{1}{2}(4x - 10) = -\frac{1}{3}(7 - 12 - x)$ 11. $4 - 3a = 7 - 2(2a + 5)$ 12. $9 - 5x = 12 - (6x + 7)$

13. $3x + 4 + 7x = 2(5x + 2)$ 14. $5x + 9 = 5x - 15$

Answers to Exercise Set 2.5

1. 3   2. 8   3. 2

4. –8   5. 0   6. 1

7. 2   8. 3   9. 1

10. 4   11. –7   12. –4

13. all real numbers   14. no solution

2.6   Ratios and Proportions

Summary

> A ratio is a quotient of two quantities. The ratio of a
> to b may be written as $a{:}b$ or $\dfrac{a}{b}$.

Example 1
A class consists of 14 males and 17 females.

<u>Answers</u>

   a. Find the ratio of females to males.   17:14 or $\dfrac{17}{14}$

   b. Find the ratio of males to females.   14:17 or $\dfrac{14}{17}$

   c. Find the ratio of males to the entire class.   14:31 or $\dfrac{14}{31}$

Example 2
Find the ratio of 8 inches to 2 feet.

We must first express with the same unit. So, 2 feet will be expressed as 2(12 inches) = 24 inches.

Thus, 8:24 or $\dfrac{8}{24}$ or $\dfrac{1}{3}$.

Summary

> A proportion is a special type of equation. It is a
> statement of equality between 2 ratios. One way of
> denoting a proportion is $a{:}b = c{:}d$ which is read "a is
> to b as c is to d." In this text we write proportions as
> $a/b = c/d$.
>
> The a and d are referred to as extremes, and the b
> and c are referred to as means of the proportion. One
> method that can be used in evaluating proportions is
> cross-multiplication.
>
> Cross-multiplication
>
> If $\dfrac{a}{b} = \dfrac{c}{d}$ then $ad = bc$.

Example 3
Solve the following for x using cross multiplication.

a. $\dfrac{x}{8} = \dfrac{3}{4}$

b. $\dfrac{3}{-x} = \dfrac{4}{6}$

Solution

a. $\dfrac{x}{8} = \dfrac{9}{4}$

$4 \cdot x = 8 \cdot 9$

$4x = 72$

$\dfrac{4x}{4} = \dfrac{72}{4}$

$x = 18$

Check: $\dfrac{x}{8} = \dfrac{9}{4}$

$\dfrac{18}{8} = \dfrac{9}{4}$

$\dfrac{9}{4} = \dfrac{9}{4}$ True

b. $\dfrac{3}{-x} = \dfrac{4}{16}$

$3 \cdot 16 = -x \cdot 4$

$48 = -4x$

$\dfrac{48}{-4} = \dfrac{-4x}{-4}$

$-12 = x$

Check: $\dfrac{3}{-x} = \dfrac{4}{16}$

$\dfrac{3}{-(-12)} = \dfrac{4}{16}$

$\dfrac{3}{12} = \dfrac{4}{16}$

$\dfrac{1}{4} = \dfrac{1}{4}$ True

Summary

To Solve Problems Using Proportions

1. Understand the problem.

2. Translate the problem into mathematical language.

   a. First, represent the unknown quantity by a variable (a letter).

   b. Second, set up the proportion by listing the left side of the equal sign, and the unknown and the other given quantity on the right side of the equal sign. When setting up the right side of the proportion, the same respective quantities should occupy the same respective positions on the left and the right. For example, an acceptable proportion might be

   Given ratio $\left\{ \dfrac{\text{miles}}{\text{hour}} = \dfrac{\text{miles}}{\text{hour}} \right.$

3. Carry out the mathematical calculations necessary to solve the problem.

   a. Once the proportion is correctly written, drop the units and cross-multiply.

   b. Solve the resulting equation.

4. Check the answer obtained in step 3.

5. Make sure you have answered the question.

Note: The two ratios must have the same units. For example, if one ratio is given in miles/hour and the other ratio is given in feel/hour, one of the ratios must be changed before setting up the proportion.

Example 4

During the first 560 miles of their vacation trip, the Smith's auto used 18 gallons of gasoline. At this rate, how many gallons to the nearest tenth of a gallon will be needed to finish the remaining 420 miles of the trip?

Solution

(step 1) The given ratio is 560 miles for 18 gallons of gas. The unknown quantity is the amount of gas needed to finish the trip.

(step 2)   Let x = number of gallons
             Given ratio is 560 miles to 18 gallons

   Proportion is $\dfrac{560 \text{ miles}}{18 \text{ gallons}} = \dfrac{420 \text{ miles}}{x \text{ gallons}}$

(step 3) $\dfrac{560}{18} = \dfrac{420}{x}$

$560x = 18 \cdot 420$

$560x = 7560$

$\dfrac{560x}{560} = \dfrac{7560}{560}$

$x = 13.5$

(step 4)   A check shows that $(18)(13.5) = 423$ which is close to 420, due to round off.

(step 5)   13.5 gallons are needed to complete the remaining 420 miles of the trip.

Example 5
A recipe calls for 2 lb of beef to prepare 12 servings. If we have 12 oz of beef available, how many servings will that make?

Solution
Let x = number of servings with 12 oz available
First convert 2 lb to oz
2 lb = 2(16 oz) = 32 oz
given ratio is 32 oz to 12 servings
Ratio with unknown is 12 oz to x servings

$\dfrac{32 \text{ oz}}{12 \text{ servings}} = \dfrac{12 \text{ oz}}{x \text{ servings}}$

$\dfrac{32}{12} = \dfrac{12}{x}$

$32x = 12 \cdot 12$

$32x = 144$

$\dfrac{32x}{32} = \dfrac{144}{32}$

$x = 4.5$

12 ounces of beef will make 4.5 servings.

Summary

Proportions can be used to convert from one quantity to another. For example, you can use proportion to convert a measurement in feet to a measurement in meters, or to convert from U.S. dollars to Mexican pesos.

Example 6
Convert 300 dollars to British pounds if the exchange rate is $1.61 equals one British pound. Round your answer to the nearest hundredth.

Solution
Let x = number of British pounds equal to 300 dollars
Given ratio is 1 British pound to 1.61 dollars
Ratio with unknown quantity is x British pounds to 300 dollars

$$\frac{1.61 \text{ dollars}}{1 \text{ British pound}} = \frac{300 \text{ dollars}}{x \text{ British pounds}}$$

$$\frac{1.61}{1} = \frac{300}{x}$$

$$1.61x = 300 \cdot 1$$

$$1.61x = 300$$

$$\frac{1.61x}{1.61} = \frac{300}{1.61}$$

$$x \approx 186.34$$

300 dollars is equivalent to 186.34 British pounds.

Summary

> Proportions can be used to solve problems in geometry. If two figures are **similar**, then their corresponding sides are in proportion.

Example 7
The two triangles below are similar.

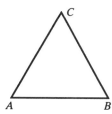

The following measurements are known $AB = 8$ inches, $XY = 5$ inches, and $AC = 5$ inches. Find the length of XZ.

Solution
Let x = Length of XZ
Known ratio is AB to XY, which is 8 to 5.
Ratio with unknown quantity is AC to XZ, which is 5 to x.

$$\frac{8 \text{ inches}}{5 \text{ inches}} = \frac{5 \text{ inches}}{x \text{ inches}}$$

$$\frac{8}{5} = \frac{5}{x}$$

$$8x = 5 \cdot 5$$

$$8x = 25$$

$$\frac{8x}{8} = \frac{25}{8}$$

$$x = 3\frac{1}{8}$$

The length of XZ is $3\frac{1}{8}$ inches.

Exercises Set 2.6

If the distribution of grades for a particular exam is 5 A's, 6 B's, 10 C's, 6 D's, and 4F's, write the ratios of the following:

   1. A's to F's   **2.** C's to B's   **3.** Total grades to B's

Determine the following ratio. Express each ratio in lowest terms:

   4. 12 yards to 20 yards   **5.** 60 minutes to 45 minutes   **6.** 80 minutes to 2 hours

   7. 25 feet to 5 yards

Solve for the variable by cross-multiplying:

   8. $\dfrac{3}{x} = \dfrac{12}{40}$ **9.** $\dfrac{x}{11} = \dfrac{40}{10}$ **10.** $\dfrac{15}{48} = \dfrac{-x}{16}$

   11. $\dfrac{15}{4} = \dfrac{30}{-x}$

Write a proportion that can be used to solve the problem then solve the problem.

   12. If 100 lb of pesticide will cover 50 acres of potatoes, how much pesticide is needed for 135 acres of potatoes?

   13. A homeowner in Florida pays $360 in property taxes on his home, which is assessed at $80,000. His neighbor's house is assessed at $145,000. How much property tax will his neighbor pay?

   14. A recipe calls for 3 lb of flour to make a cake that will serve 16 persons. If only 20 oz of flour is available to make a cake, how many persons will that cake serve?

15. A meter is equal to 39.37 inches. How many meters is equivalent to 90 inches? (Round to the nearest tenth of a meter.)

16. Find the unknown length of the two similar triangles below:
(Triangle *ABC* is similar to *XYZ*.)

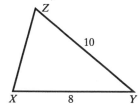

17. Find the unknown length of the two similar triangles below:
(Triangle *DEF* is similar to triangle *UVW*.)

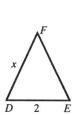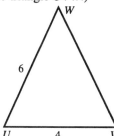

Answers to Exercise Set 2.6

1. 5:4	**2.** 10:6 or 5:3	**3.** 31:6
4. 3:5	**5.** 4:3	**6.** 2:3
7. 5:3	**8.** 10	**9.** 44
10. −5	**11.** −8	**12.** 270 pounds
13. $652.50	**14.** $6\frac{2}{3}$ persons	**15.** 2.3 meters
16. 5	**17.** 3	

Chapter 2 Practice Test

Simplify

1. −2(3x + 5)   **2.** −(x − y + 5)

Simplify when possible.

3. 3x − 5 + 7x − 6   **4.** 3a − 7a + 4b   **5.** 2(x − 3) + 3x + 2

6. −(x − 3) − x + 3

Solve each equation:

7. $3x + 5 = 11$

8. $\dfrac{3}{4}x = -9$

9. $3x - 2 = 5x + 8$

10. $3(2x - 5) = 8x - 9$

11. $11 - 4x = 2x + 8$

12. $6 - 2(5x - 8) = 3(3x - 4)$

13. $6x - 3(2 - 3x) = 4(2x - 7)$

14. $5x - 6 = 5(x - 1) - 1$

15. $\dfrac{-x}{7} = \dfrac{25}{10}$

16. $2x - 7 = 13$

17. $-2(3x + 2) = 3x + 5$

18. $3(x - 2) + 5x = 8x - 6$

19. The following triangles are similar. *ABC* is similar to *XYZ*. Find the length of the unknown side.

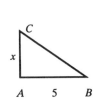 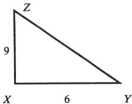

20. If a car travels 240 miles on 11 gallons of gasoline, how far would it travel on 7 gallons of gasoline? (Round to the nearest tenth of a mile.)

Answers to Chapter 2 Practice Test

1. $-6x - 10$

2. $-x + y - 5$

3. $10x - 11$

4. $-4a + 4b$

5. $5x - 4$

6. $-2x + 6$

7. 2

8. -12

9. -5

10. -3

11. $\dfrac{1}{2}$

12. $\dfrac{34}{19}$

13. $-\dfrac{22}{7}$

14. all real numbers

15. -17.5

16. 10

17. -1

18. all real numbers

19. $7\dfrac{1}{2}$ or $\dfrac{15}{2}$

20. 152.7 miles

Chapter 3

3.1 Formulas

Summary

> The **simple interest** formula is
> interest = principal $\times$ rate $\times$ time or $i = prt$.

Example 1
Stan borrows $1000 from a bank for 2 years at a rate of 14% simple interest per year. How much will Stan owe the bank after 2 years?

Solution
Use $i = prt$

Here, $p = 1000$, $i = 14\%$ or 0.14 and $t = 2$.
So, $i = 1000(0.14)(2) = \$280$
The total amount he owes the bank is $1000 + $280 or $1280.

Summary

> 1. The **perimeter** of a figure is the sum of the lengths of the sides of a figure and is measured in units such as inches, centimeters, feet, and so on.
>
> 2. The **area** of a figure is the number of square units contained inside the figure and is measured in cm^2, in^2, ft^2, and so on.
>
> 3. The table on page 164 of your textbook lists all the formulas for perimeter and area of common geometric figures such as squares, rectangles, parallelograms, trapezoids, triangles, and circles.

Example 2
Find the total length of base moulding needed along the base of the four walls of a bedroom which measures 16 feet by 10 feet.

Solution
Assume the bedroom is rectangular, and perimeter is $2l + 2w$.
That is, $P = 2(16) + 2(10)$
$\qquad P = 32 + 20 = 52$ feet

Example 3

A triangle has an area of 64 ft^2 and a height of 8 ft. What is the length of the base of the triangle?

Solution

$$A = \frac{1}{2}bh$$

$$64 = \frac{1}{2}(b)(8) \quad \text{Substitutions}$$

$$64 = \frac{1}{2}(8)b$$

$$64 = 4b$$

$$\frac{64}{4} = \frac{4b}{4}$$

$$16 = b$$

The base is 16 ft.

Example 4

The perimeter of a rectangle is 48 cm. If the width is 8 cm, find the length.

Solution

$$P = 2l + 2w$$

$$48 = 2l + 2(8) \quad \text{Substitutions}$$

$$48 = 2l + 16$$

$$48 - 16 = 2l + 16 - 16$$

$$\frac{32}{2} = \frac{2l}{2}$$

$$16 = l$$

The length is 16 cm.

Summary

> 1. The **circumference**, C, of a circle is the perimeter of the curve that forms the circle.
>
> 2. The **radius**, r, of a circle is the line segment from the center of a circle to any point on the circle.
>
> 3. A **diameter** of a circle is a line segment through the center whose endpoints both lie on the circle.

Example 5

The diameter of a pizza pie is 14 inches. What is the area and circumference of the pizza?

Solution

Since $A = \pi r^2$ and $r = \dfrac{1}{2}d$, $r = \dfrac{1}{2}(14) = 7$ inches.

Now, using 3.14 for π, we have $A = \pi(7)^2$ or $(3.14)(7)^2 = 3.14(49) = 153.86$ inches.

$C = 2\pi r$ or $C = \pi d$
We can use either formula
Using $C = 2\pi r$ we have $C = 2(3.14)(7) = (14)(3.14) = 43.96$ inches. (Note the units)

Or $C = \pi d = 3.14(14) = 43.96$ inches.

Example 6

The volume of a right circular cylinder is 706.5 in^3. What is the radius of the cylinder if the height is 9 inches.

Solution
Use $V = \pi r^2 h$
$706.5 = (3.14)r^2(9)$

$\dfrac{706.5}{28.26} = \dfrac{28.26 r^2}{28.26}$

$25 = r^2$
If $r^2 = 25$ then r must be 5 since $5^2 = 25$.
Thus, $r = 5$ inches.

Example 7
Solve $V = lwh$ for w.

Solution

1. Treat all letters as constant <u>except</u> for w.

2. Solve for w by isolating it on one side of the equation.
   $V = lwh$
   $\dfrac{V}{lh} = \dfrac{lhw}{lh}$ Rearrange terms and divide both sides by lh.
   $\dfrac{V}{lh} = w$ Simplify.

Example 8

Solve $F = \dfrac{9}{5}C + 32$ for C.

Solution

$$5F = 5\left(\dfrac{9}{5}C + 32\right)$$
$$5F = 5\left(\dfrac{9}{5}C\right) + 5(32)$$
$$5F = 9C + 160$$
$$5F - 160 = 9C + 160 - 160$$
$$\dfrac{5F - 160}{9} = \dfrac{9C}{9}$$
$$\dfrac{5}{9}F - \dfrac{160}{9} = C$$

Example 9

Solve $A = \dfrac{m + d}{2}$ for m.

Solution

$$2(A) = 2\left(\dfrac{m + d}{2}\right) \quad \text{Multiply both sides by 2}$$
$$2A = m + d$$
$$2A - d = m + d - d$$
$$2A - d = m$$

Exercise Set 3.1

Use the formula to find the value of the indicated variable.

1. $A = \dfrac{1}{2}h(b + c)$; find A when $h = 6, b = 3, c = 5$.

2. $P = 2l + 2w$; find l when $P = 40$ and $w = 16$.

3. $Z = \dfrac{x - a}{S}$; find z when $x = 130, a = 100$ and $s = 15$.

Solve each equation for y. Then find the value of y for the given value of x:

4. $3x - 2y = 6$, when $x = 3$

5. $2x - y = 10$, when $x = -4$

6. $3x - 4y = 12$, when $x = 5$

Solve for the variable indicated:

7. $I = prt$ for r

8. $ax + by = c$ for x

9. $V = \dfrac{1}{3}\pi r^2 h$ for h

10. $A = \dfrac{x + y + z}{4}$ for z

11. $P = 2l + 2w$ for l

12. $V = lwh$ for w

13. $Z = \dfrac{x - m}{5}$ for m

14. If $C = \dfrac{5}{9}(F - 32)$, find C if $F = 95°$.

15. Joe left $2000 in a bank which paid $4\dfrac{1}{2}\%$ simpla interest.
When he withdrew his money, he received $360.00 of interest.
How long did he leave his money in the account?

16. A vacant lot is in the shape of a trapezoid. If the height of the trapezoid is 12 meters, one base is 15 meters and the area is 126 square meters, find the length of the other base.

17. If the circumference of a car tire is 87.92 inches, how large is the diameter of the tire?

18. If a soup can measures 3 inches in diameter and is 7 inches high, what is its volume?

19. Find the area of a triangle whose base is 5 inches and whose height is 9 inches.

20. If the sum, S, of the first n even numbers is $S = n^2 + n$, find the sum of the first 20 even numbers.

Answers to Exercise Set 3.1

1. 24

2. $l = 4$

3. $z = 2$

4. $y = \dfrac{3}{2}x - 3,\ 1\dfrac{1}{2}$

5. $2x - 10 = y, -18$

6. $\dfrac{3}{4}x - 3 = y,\ \dfrac{3}{4}$

7. $\dfrac{I}{pt} = r$

8. $x = -\dfrac{by}{a} + \dfrac{c}{a}$

9. $\dfrac{3V}{\pi r^2} = h$

10. $4a - x - y = z$

11. $\dfrac{P - 2w}{2} = l$

12. $\dfrac{V}{lh} = w$

13. $m = x - 5Z$

14. $C = 35°$

15. 4 years

16. 6 meters

17. 28 inches

18. 49.455 square inches

19. 22.5 square inches

20. $S = 420$

3.2 Changing Application Problems into Equations

Summary

> Statements can be represented as algebraic expressions.

The following verbal statements have been translated to algebraic statements.

<u>Verbal</u>	<u>Algebraic</u>
a number increased by 6	$x + 6$
8 more than a number	$x + 8$
5 less than a number	$x - 5$
a number decreased by 7	$x - 7$
3 times a number	$3x$
four fifths of a number	$\dfrac{4}{5}x$
a number divided by 4	$\dfrac{x}{4}$
7 more than 3 times a number	$3x + 7$
6 less than 4 times a number	$4x - 6$
2 times the difference of a number and 5	$2(x - 5)$

Example 1
Express each phrase as an algebraic expression.

a. The area increased by 8 square centimeters $\qquad A + 8$

b. 5 more than twice the perimeter $\qquad 2p + 5$

    c. Twice the sum of the length and 5 $2(l + 5)$

    d. 5 more than twice the volume $5 + 2V \text{ or } 2V + 5$

Example 2
Write the three different verbal statements to represent the following expression.
$7x + 3$

Solution

    1. Three more than seven times a number

    2. Seven times a number increased by 3

    3. The sum of seven times a number and 3

Note: Some students may incorrectly interpret this last verbal statement as $7(x + 3)$. However, this mathematical statement should be interpreted as the product of 7 and the sum of a number and 3.

Example 3
Write a verbal statement to represent each expression. There can be more than one answer.

    a. $5x - 3$ Two less than 5 times a number

    b. $5(x - 2)$ Five times the difference of a number and 2

Example 4
Represent the following statement mathematically:
The number of violent crimes decreased by 9%.

Solution
Let c = previous number of violent crimes
Then $c - 0.09c$ represents the present number of violent crimes.
This statement is of the general form:
Original amount decreased by a percentage of the original amount equals the new amount.
Or $A - r\%A$ = new amount

Example 5
The population of a city increased by 4%. Represent this statement mathematically.

Solution
$p + 4\%p$ = new population
$p + 0.04p$ = new population

Common error: Some students write the answer to this question as $p + 0.04$. It is important to realize that a percent of a quantity must always be a percent multiplied by some letter or number.

Example 6
A 25 foot board is cut into 2 pieces. If x represents one piece, then express the length of the other piece in terms of x.

Solution
Notice the pattern which follows:

<u>1st piece</u>   <u>2nd piece</u>

5                              $25 - 5 = 20$
8                              $25 - 8 = 17$
x $25 - x$

Example 7
A 15-foot board is cut into two pieces. One piece is twice as long as the other piece. Find an algebraic expression for the length of the 2 pieces.

Solution
Smaller piece is x; the larger piece is $2x$

Note: We also know that $x + 2x = 15$.

Example 8
Express each of the following as an algebraic expression.   <u>Answers</u>

   a. Cost of purchasing y items at \$5 each $5y$

   b. 15% commission on x dollars in sales $0.15x$

   c. Cost of producing y items at 35¢ each and x items at 54¢ each $0.35y + 0.54x$

   d. Distance traveled in t hours at a rate of 65 miles per hour $65t$

   e. The numbers of ml of intravenous fluid in x hours at a rate of
      150 ml of fluid per hour $150x$

   f. The number of seconds in a minutes and b seconds $60a + b$

   g. The sum of two consecutive integers if the first integer is n
      Example: $3 + 4$
           $5 + 6$
           $7 + 8$
        $100 + 101$ so $(n) + (n + 1)$ or $2n + 1$

   h. The product of two consecutive **odd** integers if the first **odd** integer is x.
      Example:
        $3 \cdot 5$
        $5 \cdot 7$
      $101 \cdot 103$
      $x \cdot (x + 2)$ or $x^2 + 2x$

Summary

The word "is" in a verbal problem means "is equal
to" and is represented by an equal sign.

In Examples 9–11, find the equation.

Example 9
The sum of three times a number and four is 7.
Equation: $3x + 4 = 7$

Example 10
The product of 2 consecutive even integers is 80.

Equation
$n(n + 2) = 80$

Example 11
A number increased by 20% is 12.

Equation
$n + 0.20n = 12$

Example 12
Write a verbal statement for $y + 3(y - 5) = 7$.

Solution
The sum of a number y and 3 times the difference of y and 5 is seven.

Example 13
Write an equation for the cost of a skirt at 25% off sale is $65.

Solution
$x - 0.25x = 65$

Example 14
Write an equation for "One car travels 5 times as far as a second car. The total distance traveled is 720 miles."

Solution
$5x + x = 720$

Exercise Set 3.2

Write as an algebraic expression.

1. 5 times a number

2. 7 less than twice a number

3. 18% of a number, y

4. $8\frac{1}{2}\%$ interest rate on p dollars

5. Five less than 4 times a number

6. Four times the sum of a number and 9

7. The cost of renting y video cassettes at a rate of $2.25 per video cassette.

Express as a verbal statement.

8. $5 - y$ 　　　　　　**9.** $3x + 2$ 　　　　　　**10.** $4(x - 5)$

Select a variable to represent one quantity. State what that variable represents. Then express the second quantity in terms of the variable.

11. John's salary is $150 more than Harry's salary.

12. The cost of an item and the cost increased by a $6\frac{1}{2}\%$ sales tax.

13. The federal deficit and the federal deficit reduced by 15%.

14. The world population and the world population increased by 7%.

Express as an equation.

15. One board is 3 times as long as a second board. The sum of their lengths is 20.

16. The cost of a stereo plus 7% tax is $800.

17. One train travels 5 miles less than 3 times another train. The total distance traveled by both trains is 500 miles.

18. The product of two consecutive integers is 132.

19. The sum of two consecutive even integers is 70.

20. The total cost of a company if $500 per month plus $35 for each item produced. The total cost this month is $1200.

21. The sum of twice a number and that number increased by 3 is 8.

22. The cost of traveling x miles at 29 cents per mile is $24.90.

Answers to Exercise Set 3.2

1. $5x$ 　　　　　　**2.** $2x - 7$ 　　　　　　**3.** $0.18y$

4. $0.085p$ 　　　　　　**5.** $4x - 5$ 　　　　　　**6.** $4(x + 9)$

7. $2.25y$ 　　　　　　**8.** The number 5 decreased by y

9. Three times a number increased by 2.

10. The product of 4 and 5 less than a number.

11. x = Harry's salary; $150 + x$ = John's salary

12. x = cost of an item; $x + 0.065\,x$ = cost of the item plus sales tax

13. d = federal deficit; $d - 0.15d$

14. p = world's population; $p + 0.07p$

15. x = length of short board; $x + 3x = 20$

16. x = cost of stereo before tax; $x + 0.07x = 800$

17. d = distance traveled by one train; $3d - 5$ = distance traveled by the second train; $d + (3d - 5) = 500$

18. n = first integer; $n(n + 1) = 132$

19. n = first even integer; $n + (n + 2) = 70$

20. x = number of items; $500 + 35x = 1200$

21. x = a number; $2x + (x + 3) = 8$

22. x = number of miles driven; $0.29x = 24.90$

3.3   Solving Application Problems

Summary

Problem-Solving Procedure for Solving Applications

1. **Understand the problem.**
   Identify the quantity or quantities you are being asked to find.

2. **Translate the problem into mathematical language (express the problem as an equation).**

   a.   Choose a variable to represent one quantity, *and write down exactly what it represents.* Represent any other quantity to be found in terms of this variable.

   b.   Using the information from step a), write an equation that represents the application.

3. **Carry out the mathematical calculations (solve the equation).**

4. **Check the answer (using the *original* application).**

5. **Answer the questions asked.**

Example 1
Three is added to 5 times a number and the result is 48. Find the number.

Solution
Let x = unknown number
Write the expression
"Three is added to 5 times a number"
$3 + 5x$
"The result is 48" gives the equation
$3 + 5x = 48$
Solve the equation.
$$3 + 5x = 48$$
$$3 - 3 + 5x = 48 - 3$$
$$5x = 45$$
$$\frac{5x}{5} = \frac{45}{5}$$
$$x = 9$$

Answer the question.
"The number is 9."

Check the solution
$$3 + 5(9) = 48$$
$$3 + 45 = 48$$
$$48 = 48 \text{ True}$$

Example 2
The sum of two consecutive odd numbers is 96. Find the numbers.

Solution
Let x = first odd consecutive number. Then $x + 2$ is the next consecutive odd number.

Write the equation
Sum of two consecutive odd numbers is 96.
$x + (x + 2) = 96$

Solve the equation.
$$x + (x + 2) = 96$$
$$2x + 2 = 96$$
$$2x + 2 - 2 = 96 - 2$$
$$2x = 94$$
$$\frac{2x}{2} = \frac{94}{2}$$
$$x = 47$$

Answer the question.
$x = 47, x + 2 = 49$

Check the solution.
$47 + 49 = 96$

$\quad 96 = 96$ True

Example 3
The population of a town increased by 8% one year and reached a new population of 5400. What was the population of the town before the increase?

Solution
Let x = population before increase
Write the equation.
Population + 8% of population = new population
$x + 0.08x = 5400$

Solve the equation.
$1x + 0.08x = 5400$

$\quad 1.08x = 5400$

$\quad \dfrac{1.08x}{1.08} = \dfrac{5400}{1.08}$

$\quad\quad x = 5000$

The population of the town was 5000 people.

Example 4
A collection of nickels, dimes and quarters is worth \$3.75. There are twice as many nickels as quarters and 4 times as many dimes as quarters. How many coins of each type are in the collection?

Solution
Let x = number of quarters
Express the number of nickels and dimes in terms of the number of quarters.
x = number of quarters
$2x$ = number of nickels
$4x$ = number of dimes
Write the equation
(numbers of nickels) · value of a nickel + (number of dimes) · value of a dime + (number of quarters)
$\quad\quad$ · x value of a quarter = 3.75
Thus, $2x(0.05) + 4x(0.10) + x(0.25) = 3.75$

Solve the equation
$0.1x + 0.4x + 0.25x = 3.75$

$(0.1 + 0.4 + 0.25)x = 3.75$

$\quad\quad 0.75x = 3.75$

$\quad\quad x = \dfrac{3.75}{0.75} = 5$

Answer the question.

There are 5 quarters, $2(5) = 20$ nickels, and $4(5) = 20$ dimes.

Check the solution
$$\$5(0.25) + \$10(0.05) + \$20(0.10) = \$3.75$$
$$\$1.25 + \$0.50 + \$2.00 = \$3.75$$
$$\$3.75 = \$3.75 \text{ True}$$

Example 5
A 16-oz container of mouthwash contains 2.4 ounces of hydrogen peroxide. Find the percent by volume of hydrogen peroxide in the mouthwash.

Solution
Let x = percent of peroxide

Write the equation.
Use: total volume $\times$ percent of hydrogen peroxide = amount of hydrogen peroxide.

So, $16 \cdot x = 2.4$
Or $16x = 2.4$

Solve the equation.
$$\frac{16x}{16} = \frac{2.4}{16}$$
$$x = 0.15 \text{ or } 15\%$$

Answer the question.
The percent of hydrogen peroxide is 15%.

Check the solution
$$16(0.15) = 2.4$$
$$2.4 = 2.4 \text{ True}$$

Example 6
A taxi cab driver charges $1.10 for the first quarter mile and 40 cents for each quarter mile thereafter. If Audra has $10.00 to spend on cab fare, how far can she travel in this cab?

Solution
Let x = number of *quarter* miles the cab will be driven since the rates are expressed in terms of quarter miles.

Write the equation.
If x = number of quarter miles driven then $x - 1$ = number of quarter miles driven *after* the first quarter mile.

Cab fare $= 1.10 + 0.40(x - 1)$
$$10 = 1.10 + 0.40(x - 1)$$

Solve the equation.
$$10 = 1.10 + 0.40(x - 1)$$
$$10 = 1.10 + 0.40x - 0.40$$
$$10 = 0.70 + 0.40x$$
$$10 - 0.7 = 0.40x$$
$$9.3 = 0.40x$$
$$9.3 = 0.40x$$
$$\frac{9.3}{0.4} = x$$
$$23.25 = x$$

Answer the question.

Audra can be driven 23.25 quarter miles or $\dfrac{23.25}{4}$ or about 5.8 miles.

Check the solution.
$$\$1.10 + \$(23.25 - 1)(0.40) = \$10.00$$
$$\$1.10 + \$8.90 = \$10.00$$
$$\$10.00 = \$10.00 \text{ True}$$

Exercise Set 3.3

Set up an algebraic equation that can be used to solve the problem. Solve the equation and answer the question asked.

1. The sum of two consecutive even integers is 86. What are the integers?

2. The sum of two numbers is 39. One number is one less than three times the other number. What are the numbers?

3. The sum of three consecutive odd integers is 93. What are the integers?

4. Mr. Fernandez has $4.50 in quarters and dimes. How many quarters and how many dimes does he have if the total number of coins is 27?

5. A 32-ounce acid solution has 3.4 ounces of acid. Find the percentage (by volume) of acid in the solution.

6. A small town has a population of 4000. If the population is increasing by 150 each year, how long will it take the town to reach a population of 5800?

7. The cost of renting a car is $25 per day plus 40 cents per mile. Find the maximum number of miles the rental car can be driven in one day if $125 is to spend on renting the car.

8. A machine originally costs $10,000. If this cost depreciates by the same amount each year for 12 years, the machine will be worth $6640 after this time. Find the amount of depreciation per year for this machine.

9. Allen purchased a pair of slacks at a 20% off sale for a price of $28.79 (before sales tax is added). What was the original price of the slacks?

10. A plumber charges a flat rate of $45 per hour. A second plumber charges a one time fee of $75.00 plus $30.00 per hour. What are the numbers of hours that each plumber can work on a particular job if the cost for completing the job is the same for each plumber? What is the cost?

11. While on vacation in another state, Ms. Walsh purchased some film, postcards, and other miscellaneous items. The bill before the tax was $26.00. The bill after tax was $27.82. Find the local sales tax rate.

12. A soccer coach purchased 9 jerseys at $12.75 and an equal number of pairs of soccer cleats. If the total bill (before sales tax is added) is $258.66, what is the price of a pair of soccer cleats?

Answers for Exercise Set 3.3

1. 42, 44	**2.** 29, 10	**3.** 29, 31, 33
4. 12 quarters, 15 dimes	**5.** 10.6	**6.** 12 years
7. 250 miles	**8.** $280	**9.** $35.99
10. 5 hours, cost is $225	**11.** 7%	**12.** $15.99

3.4 Geometric Problems

Summary

Important Geometric Formulas to recall:

1. Perimeter of a rectangle:
   $P = 2l + 2w$

2. Area of a rectangle:
   $A = l \cdot w$

3. Sum of the interior angles of a triangle is 180°

4. Sum of the interior angles of a quadrilateral is 360°

5. Consult the Appendix of your text for other formulas.

Example 1
The perimeter of a rectangle garden is 38 feet. The length is 5 feet longer than the width. Find the dimensions of the garden.

Solution
Let x = width. Then length is $5 + x$.
Write the equation.
$P = 2l + 2w$
$38 = 2(5 + x) + 2x$

Solve the equation.
$$38 = 2(5) + 2x + 2x$$
$$38 = 10 + 4x$$
$$38 - 10 = 10 - 10 + 4x$$
$$28 = 4x$$
$$\frac{28}{4} = \frac{4x}{4}$$
$$7 = x$$

Answer the question.
7 feet is the width
$7 + 5 = 12$ feet is the length

Check the solution.
$$2(7) + 2(12) = 38$$
$$14 + 24 = 38$$
$$38 = 38 \text{ True}$$

Example 2
Find the three angles of a triangle if one angle is 20 ° more than the smallest angle and the largest angle is 10 more than 4 times the smallest angle.

Solution

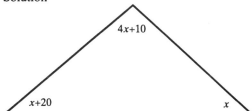

Let x = smallest angle. Then $x + 20$ is the middle angle and $4x + 10$ is the largest angle. Since the sum of the interior angles of a triangle is 180°, we have $x + (x + 20) + (4x + 10) = 180$.

Solve the equation.
$$x + (x + 20) + (4x + 10) = 180$$
$$6x + 30 = 180$$
$$6x + 30 - 30 = 180 - 30$$
$$6x = 150$$
$$\frac{6x}{6} = \frac{150}{6}$$

Answer the question.
$x = 25°$, $x + 20 = 45°$, and $4x + 10 = 110°$

The 3 angles are 25°, 45°, 110°.

Check the solution
$$25° + 45° + 110° = 180°$$
$$180° = 180°$$

Example 3
If the two larger angles in a parallelogram are 30° less than twice the smaller angles, find the measure of each angle.

Solution
FACT: In a parallelogram, opposite angles have the same measure.

Let x = measure of the smaller angle.
Then, $2x - 30$ is the measure of the larger angle.

FACT: Since a parallelogram is a quadrilateral, the sum of the measures of all four angles is 560°.

Write the equation.
$$x + x + (2x - 30) + (2x - 30) = 300$$

Solve the equation.
$$2x + 4x - 60 = 360$$
$$6x - 60 = 360$$
$$6x - 60 + 60 = 360 + 60$$
$$6x = 420$$
$$x = 70$$

Answer the question
The measure of the smaller angle is 70° while the larger angle measures $2(70) - 30 = 110°$.

Example 4
A bookshelf (shown) is to have three shelves, including the top. The height of the bookshelf is to be 1 foot less than twice the width. Find the dimensions of the bookshelf if 33 feet are available.

Solution
Let x = width of bookshelf. then, $2x - 1$ is the height of the bookshelf.

Write the equation.
(3 shelves) + (2 sides) = total lumber available
$3x + 2(2x - 1) = 33$

Solve the equation.
$$3x + 4x - 2 = 33$$
$$7x - 2 = 33$$
$$7x = 35$$
$$x = 5$$

Answer the question.
The width of the shelf is 5 feet. The height of the shelf is $2(5) - 1 = 9$ feet.

Check the solution.
$$3(5) + 2(9) = 33$$
$$15 + 18 = 33$$
$$33 = 33$$

Exercise Set 3.4

Write an algebraic equation for each of the following problems. Solve the equation and answer the question.

1. The perimeter of a room is 52 feet. The length of the room is four feet less than twice the width. Find the dimensions.

2. One angle of a right triangle is twice as large as the smallest angle. Find the measure of the three angles of this triangle.

3. The perimeter of an isosceles triangle is 23 meters. Find the length of all three sides if the two sides having the same length are 1 meter less than twice the length of the base.

4. Find the cost of installing baseboard along the base of the walls of a rectangular room which measures 12 feet by 18 feet if the cost of the baseboard is 30¢ per linear foot.

5. In a triangle, one angle is twice as large as another angle and the third angle is 40° larger than the smallest angle. Find the measure of all three angles.

6. In a parallelogram, the measure of the two largest angles is 20° more than three times the measure of the two smaller angles. Find the measure of all the angles.

7. A bookshelf is to consists of five shelves (including the top) and two sides which measure 1.5 times the length of each shelf. If 64 board feet of lumber are available, what are the dimensions of the bookshelf?

8. Moses wants a garden whose length is 4 meters longer than its width. The perimeter of the garden is to be 70 ft. What are the dimensions of the garden?

9. If angles A and B are supplementary, and angle B is 10° more than angle A, find the measure of each angle.

10. Angles A and B are complementary. Angle B is 9° more than twice the measure of angle A. Find the measure of each angle.

11. If triangle ABC is equilateral and its perimeter is 25.5 ft, find the length of each side.

12. The volume, V, of a box is given by $V = lwh$. What happens to the volume if the length, width and height are all tripled?

13. In the equation, $A = lw$, what happens to the area is the length is quadrupled and the width is halved?

14. In the formula $A = s^2$, what happens to the area if the length of a side, s, is tripled?

Answers to Exercise Set 3.4

1. 10 feet by 16 feet	2. 30°, 60°, 90°	3. 5 feet, 9 feet, 9 feet
4. $18.00	5. 35°, 70°, 75°	6. 40°, 40°, 140°, 140°
7. 8 feet by 12 feet	8. $w = 15.5$ feet, $l = 19.5$ feet	9. $\angle A = 85°$, $\angle B = 95°$
10. $\angle A = 27°$, $\angle B = 63°$	11. 8.5 feet	

12. The volume becomes 27 times as great.

13. The area becomes twice as great.

14. The area becomes 9 times as great.

3.5 Motion and Mixture Problems

Summary

Motion Problems

Amount = rate · time

Example 1

A swimming pool is being emptied at the rate of 18 gallons per minute. If the pool contained 5400 gallons of water, how long will it take to empty the pool?

Solution
Let x = time in minutes

Amount = rate · time

$$5400 = 18 \cdot x$$

$$\frac{5400}{18} = x$$

$$300 = x$$

It takes 300 minutes (5 hours) to empty the pool.

Summary

When the "amount" in the rate formula is "distance," we often refer to the formula as the **distance formula.**

Distance = rate · time or $d = r \cdot t$

Example 2

A trip takes an automobile a total of 8 hours. If the trip is a distance 440 miles, what is the average rate of the car on the trip?

Solution
Let x = average rate of the car

Distance = rate · time

$$440 = x \cdot 8$$

$$\frac{440}{80} = x$$

$$55 = x$$

The average rate of the car is 55 miles per hour.

Example 3
Two cars, one traveling 10 mph faster than the second car, start at the same time and travel in opposite directions. In 3 hours, they are 300 miles apart. Find the rate of each car.

Solution
Let x = rate of one car
then $x + 10$ = rate of the faster car

	rate	time	= distance
first car	x	3	$3x$
faster car	$x + 10$	3	$3(x + 10)$

Distance of first car + Distance of faster car = 300

$$
\begin{aligned}
3x \quad + \quad 3(x+10) \quad &= 300 \\
3x \quad + \quad 3x + 30 \quad &= 300 \\
6x + 30 \quad &= 300 \\
6x + 30 - 30 \quad &= 300 - 30 \\
6x \quad &= 270 \\
\frac{6x}{6} \quad &= \frac{270}{6} \\
x \quad &= 45
\end{aligned}
$$

Thus, the first car rate is 45 mph. Faster car rate is $x + 10 = 45 + 10 = 55$ mph.

Example 4
On a 105 mile trip, a car traveled at an average speed of 45 mph and then reduced the speed to 30 mph for the remainder of the trip. The trip took a total of 150 minutes. For how long did the car travel at 30 mph?

Solution
First of all convert 150 minutes to hours.
$$
150 \text{ minutes} = (150 \text{ minutes})\left(\frac{1 \text{ hour}}{60 \text{ minutes}}\right) = 2.5 \text{ hours}
$$

Let x = travel time at 30 mph
then $2.5 - x$ = travel time at 45 mph

	rate	· time	= distance
30 mph	30	x	$30x$
45 mph	45	$2.5 - x$	$45(2.5 - x)$

(Distance at 30 mph) + (Distance at 45 mph) = (Total distance)

$$
\begin{array}{llll}
30x & + & 45(2.5 - x) & = 105 \\
30x & + & 112.5 - 45x & = 105 \\
-15x & + & 112.5 & = 105 \\
-15x & + & 112.5 - 112.5 & = 105 - 112.5 \\
& & -15x & = -7.5 \\
& & \dfrac{-15x}{-15} & = \dfrac{-7.5}{-15} \\
& & x & = 0.5
\end{array}
$$

Therefore, the car travels 0.5 hrs (30 minutes) at 30 mph.

Summary

Any problem in which two or more quantities are combined to produce a different quantity or a single quantity is separated into two or more different quantities may be considered a **mixture problem**.

Example 5
A coffee merchant wants to make 6 pounds of a blend of coffee costing $5 per pound. The blend is made using a $6 grade and a $3 grade of coffee. How many pounds of each of these grades should be used?

Solution
Let x = number of lbs of $6 grade
then $6 - x$ = number of lbs of $3 grade

Price	Number of Pounds	Values of Coffee
$6	x	$6x$
$3	$6 - x$	$3(6 - x)$
$5	6	$5 \cdot 6$

(Value of $6 coffee) + (value of $3 coffee) = (value of $5 coffee)

$$
\begin{array}{llll}
6x & + & 3(6 - x) & = 5 \cdot 6 \\
6x & + & 18 - 3x & = 30 \\
& & 3x + 18 & = 30 \\
& & 3x + 18 - 18 & = 30 - 18 \\
& & 3x & = 12 \\
& & \dfrac{3x}{3} & = \dfrac{12}{3} \\
& & x & = 4
\end{array}
$$

Thus, 4 pounds at $6 per pound and $6 - 4 = 2$ pounds at $3 per pound must be mixed.

Example 6
How many gallons of a 20% salt solution must be mixed with 6 gallons of a 30% salt solution to make a 22% salt solution?

Solution
Let x = the number of gallons of a 20% salt solution

Solution	Strength	Gallons	Amounts of Pure Salt
20%	0.20	x	$0.20x$
30%	0.30	6	$6(0.30)$
22%	0.22	$x + 6$	$0.22(x + 6)$

$$\begin{pmatrix} \text{Amount of Pure Salt} \\ \text{in 20\% solution} \end{pmatrix} + \begin{pmatrix} \text{Amount of Pure Salt} \\ \text{in 30\% solution} \end{pmatrix} = \begin{pmatrix} \text{Amount of Pure Salt} \\ \text{in 22\% solution} \end{pmatrix}$$

$$
\begin{aligned}
0.20x \quad + \quad 6(0.30) &= 0.22(x + 6) \\
0.20x \quad + \quad 1.8 &= 0.22x + 1.32 \\
0.20x - 0.22x \quad + \quad 1.8 &= 0.22x - 0.22x + 1.32 \\
-0.02x \quad + \quad 1.8 &= 1.32 \\
-0.02x \quad + \quad 1.8 - 1.8 &= 1.32 - 1.8 \\
-0.02x &= -0.48 \\
\frac{-0.02x}{-0.02} &= \frac{-0.48}{-0.02} \\
x &= 24
\end{aligned}
$$

We must mix 24 gallons of the 20% solution.

Example 7
We have a total of $80,000 to be invested. Part will be invested at 5% simple interest and the rest at 10% simple interest. How much should be invested at each rate to return a total interest of $5000 for one year?

Solution
Let x = amount of invested at 5%
Then $80,000 - x$ = amount of invested at 10%

Account	Principal	Rate	Time	Interest
5%	x	0.05	1	$0.05x$
10%	$80000 - x$	0.10	1	$0.10(8000 - x)$

$$\left(\begin{array}{c}\text{Interest from 5\%}\\ \text{account}\end{array}\right) + \left(\begin{array}{c}\text{Interest from 10\%}\\ \text{account}\end{array}\right) = (\text{Total investment})$$

$$
\begin{array}{ccl}
0.05x & + & 0.10(80,000 - x) = 5000 \\
0.05x & + & 8000 - 0.10x = 5000 \\
-0.05x & + & 8000 = 5000 \\
-0.05x & + & 8000 - 8000 = 5000 - 8000 \\
-0.05x & & = -3000 \\
\dfrac{-0.05x}{-0.05} & & = \dfrac{-3000}{-0.05} \\
& & x = 60,000
\end{array}
$$

We need to invest $60,000 at a 5% rate and $80,000 − $60,000 = $20,000 at 10% rate.

Exercise Set 3.5

Set up an equation that can be used to solve each problem. Solve the equation and answer the question. Use a calculator when you feel it is appropriate.

1. How fast must a plane travel to fly 1200 miles in 3 hours?

2. A person walked a total of 6 miles in $1\frac{1}{2}$ hours. At what rate did the person talk?

3. A patient is to receive 1200 cubic centimeters of an intravenous fluid over a period of 8 hours. What should be the intravenous flow rate?

4. A typist can type 900 words in 12 minutes. At what rate can this typist type?

5. Two cyclists start from the same point and ride in opposite directions. One cyclist rides twice as fast as the other. In three hours, they are 72 miles apart. Find the rate of each of cyclist.

6. A motorboat leaves a harbor and travels at an average rate of 8 mph toward a small island. Two hours later a cabin cruiser leaves the same island and travels at an average speed of 16 mph toward the same island. In how many hours after the cabin cruiser leaves will the cabin cruiser be alongside the motorboat?

7. A family drove to a resort at an average speed of 30 mph and later returned over the same road at an average speed of 50 mph. Find the distance to the resort if the total driving time was 8 hrs.

8. Running at an average rate of 8 meters/second, a sprinter ran to the end of a track and then jogged back at an average rate of 3 meters/second. The sprinter took $\frac{11}{12}$ of a minute to run to the end of the track and jog back. Find the length of the track.

9. A car traveling at 48 mph overtakes a cyclist who, riding at 12 mph, has a 3 hr head start. How far from the starting point does the car overtake the cyclist?

10. On a 195-mile trip, a car traveled at an average speed of 45 mph and then reduced the speed to 30 mph for the remainder of the trip. The trip took a total of 5 hours. For how long did the car travel at each speed?

11. How many ounces of pure gold that cost $400 per ounce must be mixed with 20 oz of an alloy that cost $220 an ounce to make an alloy costing $300 per ounce?

12. How many pounds of walnuts that cost $1.60 per pound must be mixed with 18 lb of cashews that cost $2.50 per pound to make a mixture that costs $1.90 per pound?

13. Find the cost per pound of a mixture of coffee made from 25 lb of coffee that costs $4.82 per pound and 40 lb of coffee that costs $3.00 per pound.

14. A farmer has some cream that is 21% butterfat and some that is 15% butterfat. How many gallons of each must be mixed to product 60 gallons of cream that is 19% butterfat?

15. A goldsmith has 10 grams of a 50% gold alloy. How many grams of pure gold should be added to the alloy to make a mixture that is 75% gold?

16. A manufacturer mixed a chemical that was 60% fire retardant with 70 lb of a chemical that was 80% fire retardant to make a mixture that is 74% fire retardant. How much of the 60% mixture was used?

17. A total of $5000 is deposited into two simple interest accounts. On one account, the annual simple interest rate is 7%, while on the second account, the annual simple interest rate is 9%. How much should be invested in each account that the total interest earned is $410?

18. A club invested a part of $10,000 in a 6.5% simple annual interest account and the remainder in a 9% annual simple interest account. The amount of interest earned for one year was $725. How much was invested in each account?

19. An investment of $4000 is made at an annual simple interest rate of 7%. How much additional money must be invested at an annual simple interest rate of 10% so that the total interest earned is 8% of the total investment?

20. An investment of $3500 is made at an annual simple interest rate of 6%. How much additional money must be invested at an annual simple interest rate of 9% so that the total interest earned is 8% of the total investment?

Answers to Exercise Set 3.5

1. 400 mph	**2.** 4 mph	
3. 150 cubic centimeters per hour	**4.** 75 words per minute	**5.** 8 mph, 16 mph
6. 2 hours	**7.** 150 miles	**8.** 120 meters
9. 48 miles	**10.** 3 hours at 45 mph 2 hours at 30 mph	
11. 16 ounces	**12.** 36 pounds	**13.** $3.70 per pound
14. 40 gallons of 21% butterfat 20 gallons of 15% butterfat		
15. 10 grams	**16.** 30 pounds	**17.** $2000 at 7% $3000 at 9%

18. $7000 at 6.5%
$3000 at 9%

19. $2000   **20.** $7000

Chapter 3 Practice Test

1. Given $m = \dfrac{a+b}{2}$, find b when m is 36 and a is 16.

2. Find H is $V = \dfrac{1}{3}BH$, $B = 240$ and $V = 960$.

3. Solve for y then find the value of y for the given value of x.
$3x - 4y = 12; x = 8$

4. Solve $P = 2l + 2w$ for l.

5. Solve $F = \dfrac{5}{9}C + 32$ for C.

6. Solve $V = \dfrac{4}{3}\pi r^2 h$.

7. Solve $2x - 5y = 10$ for y.

Write an algebraic equation for each problem. Solve the equation and then answer the question.

8. The sum of two consecutive odd integers is 164. Find the integers.

9. What is the cost of a car before tax if the total cost of the car including a 7% sales tax is $8744?

10. The first side of a triangle is 5 meters less than the second side, and the third side is twice as long as the first side. If the perimeter is 33 meters, find the lengths of all 3 sides of the triangle.

11. The cost of renting a truck is $50.00 plus $22.00 per day. If Rodriguez has $120.00 to spend on truck rental, for how many days can he rent the truck?

12. The 2 larger angles of a parallelogram are 30° more than twice the measure of the smaller angles. Find the measure of each angle of the parallelogram.

13. A photocopier is on sale for $460.00. Copies can be made at a local copy center for 4 cents a copy. How many copies can be made for the copying cost to equal the cost of the copier?

14. You have $20.00 to spend at the grocery store. Your purchases so far total $13.10. How many large bags of corn chips can you buy at $2.30 per bag with your remaining money?

15. A bricklayer can lay 200 bricks in 40 minutes. How many bricks can he lay per hour?

16. A car travels a distance of 320 miles in a total of 7 hours. If it travels part of the time at 50 mph and part of the time at 40 mph, determine the time spent traveling at the two different rates.

17. A total of \$40,000 is to be invested. Part will be invested at 8% annual simple interest rate and the remainder at 5% annual simple interest rate. How much should be invested at each rate to have a total interest returned of \$2300?

Answers to Chapter 3 Practice Test

1. $b = 56$

2. $H = 12$

3. $y = \dfrac{3}{4}x - 3,\ y = 3$

4. $l = \dfrac{P - 2w}{2}$ or $l = \dfrac{1}{2}(P - 2w)$

5. $C = \dfrac{9}{5}(F - 32)$

6. $h = \dfrac{3V}{4\pi r^2}$

7. $y = \dfrac{2}{5}x - 2$

8. 81, 83

9. \$8,200

10. 7 meters, 12 meters, 14 meters

11. 3 days, he has \$4.00 to spare

12. 50°, 50°, 130°, 130°

13. 11,500

14. 3

15. 300 bricks per hour

16. 4 hours at 50 mph
3 hours at 40 mph

17. \$10,000 at 8%
\$30,000 at 5%

Chapter 4

4.1 The Cartesian Coordinate System and Linear Equations in two Variables

Summary

1. The Cartesian Coordinate System is formed by two axes (number lines) drawn perpendicular to each other.

2. The horizontal axis is called the *x*-axis. The vertical axis is called the **y-axis**. The point of intersection of the two axes is called the **origin.**

3. To locate a point, it is necessary to know both the *x*-value and *y*-value or **coordinates** of the point.

4. A point is specified by an **ordered pair**. (a, b). In the **ordered pair** (a, b) the *x*-coordinate is listed first, and the *y*-coordinate is second.

Example 1
Plot $A(-3, 5)$, $B(2, 4)$, $C(0, 3)$, $D(3, 0)$.

Solution

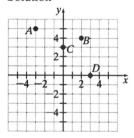

Example 2
List ordered pairs for each point shown and identify the quadrant in which they are located.

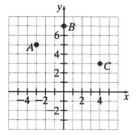

Solution
$A(-3, 5)$, $B(0, 7)$ and $C(4, 3)$
A is in quadrant II.
C is in quadrant I.
B is on the y-axis. It is not located in a specific quadrant.

Summary

> 1. A **linear equation in two variables** is an equation that can be put in the form $ax + by = c$ where a, b, c are real numbers.
>
> 2. Equations of the form $ax + by = c$ will be straight lines when graphed.
>
> 3. **A graph** is an illustration of a set of points which satisfy the equation.

Example 3
Given $y = 2x + 1$, solutions of this equation will be ordered pairs which satisfy this equation. For example, if $x = 1$, then $y = 2(1) + 1 = 3$, so, $(1, 3)$ is a solution. If $x = -2$, then $y = 2(-2) + 1 = -4 + 1 = -3$, so, $(-2, -3)$ is also a solution. If $x = 0$, then $y = 2(0) + 1 = 1$, so, $(0, 1)$ is the third solution. If we plot the 3 ordered pairs $(1, 3)$, $(-2, -3)$, and $(0, 1)$, we obtain the graph shown.

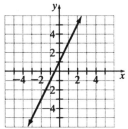

The three points lie on the same line. Countless other points also lie on this line.

Example 4
Determine which of the following ordered pairs satisfy $x - y = 1$.

   a. $(0, 1)$ **b.** $(2, 3)$ **c.** $(-3, -4)$

Solution

   a. We substitute the ordered pair $(0, 1)$ into $x - y = 1$ to obtain $0 - 1 = 1$ or $-1 = 1$ which is false. Thus, $(0, 1)$ does not satisfy the equation.

   b. Substitute $(2, 3)$ into $x - y = 1$ to obtain $2 - 3 = 1$ or $-1 = 1$ which is false. Thus, $(2, 3)$ is not a solution.

   c. Substitute $(4, 3)$ into $x - y = 1$. Thus, $4 - 3 = 1$ or $1 = 1$ which is true so $(4, 3)$ satisfies the equation.

Example 5
Determine if the three points are collinear.

    a. $(0, 1), (2, 5), (-2, -3)$ **b.** $(0, 3), (1, 2), (3, 0)$ **c.** $(0, 0), (2, 8), (3, 9)$

Solution

If three points are collinear, then that means they lie on the same line. We will plot each set of points on the grid to determine if they lie on the same line.

a.

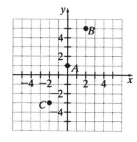

b.

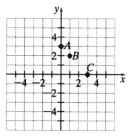

c.

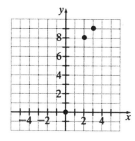

The points in a) and b) are collinear. The points in c) are **not** collinear.

Exercise Set 4.1

Indicate the quadrant in which each of the points belongs.

    1. $(3, 6)$ **2.** $(-3, 5)$ **3.** $(-3, -5)$

    4. $(100, 93)$ **5.** $(5, -100)$

    6. List the ordered pairs corresponding to the points graphed.

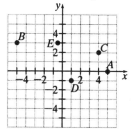

7. Plot the points. Then determine if they are collinear.

    a. (0, 3), (1, 2), (3, 0)   **b.** (1, 3), (−2, 1), (0, 5)   **c.** (−2, −8), (2, 4), (−3, −10)

8. Determine if the ordered pairs satisfy the equation.

$$y = \frac{2}{3}x - 2$$

    a. (−3, −4)   **b.** (0, −2)   **c.** (6, 0)

    d. (3, 0)   **e.** (9, 3)

9. Given $y = 2x + 1$, complete the ordered pairs so that each ordered pair satisfies the equation.
(−3, ___) (−1, ___) (0, ___) (2, ___) (3, ___)

10. a. How many points are needed to graph a linear equation?

    b. Why is it preferable to use three or more points when graphing a linear equation?

11. What does it mean when one says "three points are collinear"?

Answers to Exercise 4.1

1. I   **2.** II   **3.** III

4. I   **5.** IV

6. $A(5, 0)$, $B(−5, 3)$, $C(4, 2)$, $D(1, −1)$, $E\left(-\frac{1}{2},\ 3\right)$

7. a. yes   **b.** yes   **c.** no

8. a. yes   **b.** yes   **c.** no

    d. yes   **e.** no

9. −5, −1, 1, 5, 7

10. a. two   **b.** To catch errors

11. The 3 points all lie on the same line.

4.2   Graphing Linear Equations

Summary

Graphing Linear Equations by Plotting Points

1. Solve the linear equation for the variable y. That is, get the variable y by itself on the left side of the equal sign.

2. Select a value for the variable x. Substitute this value in the equation for x and find the corresponding value of y. Record the ordered pair (x, y).

3. Repeat step 2 with two different values of x. This will give you two additional ordered pairs.

4. Plot the three ordered pairs. The three points should be collinear. If they are not collinear, recheck your work for mistakes.

5. With a straight-edge, draw a straight line through the three points. Draw arrow heads on each end of the line segment to show that the line continues indefinitely in both directions.

Example 1
Graph $y = -2x + 5$.

Solution
Equation is already solved for y.

$$y = -2x + 5 \qquad \text{ordered pair:}$$

Let $x = -1$	$y = -2(-1) + 5 = 7$	$(-1,\ 7)$
Let $x = 0$	$y = -2(0) + 5$	$(0,\ 5)$
Let $x = 3$	$y = -2(3) + 5$	$(3,\ -1)$

Plot the points $(-1, 7)$, $(0, 5)$, $(3, -1)$ and draw the line.

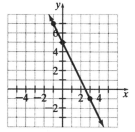

Example 2
Graph $3x - 4y = 12$.

Solution
Solve for y: $-4y = -3x + 12$

$$y = \frac{-3x + 12}{-4} = \frac{-3}{-4}x + \frac{12}{-4} = \frac{3}{4}x - 3$$

Now select values of x divisible by 4 since coefficient of x is $\frac{3}{4}$

$$y = \frac{3}{4}x - 3 \qquad\qquad \text{ordered pair:}$$

Let $x = -4$ $\quad y = \frac{3}{4}(-4) - 3 = -3 - 3 = -4$ $\quad (-4, -6)$

Let $x = 0$ $\quad y = \frac{3}{4}(0) - 3 = 0 - 3 = -3$ $\quad (0, -3)$

Let $x = 4$ $\quad y = \frac{3}{4}(4) - 3 = 3 - 3 = 0$ $\quad (4, 0)$

Plot the points and draw the line.

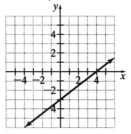

Summary

Graphing Linear Equation Using the *x*- and *y*-Intercepts

1. Find the *y*-intercept by setting *x* in the given equation equal to 0 and finding the corresponding value of *y*.

2. Find the *x*-intercept by setting *y* in the given equation equal to 0 and finding the corresponding value of *x*.

3. Determine a check point by selecting a nonzero value for *x* and finding the corresponding value of *y*.

4. Plot the *y*-intercept (where the graph crosses the *y*-axis), the *x*-intercept (where the graph crosses the *x*-axis), and the check point. The three points should be collinear. If not, recheck your work.

5. Using a straight-edge, draw a straight line through the three points. Draw arrow heads at both ends of the line to show that the line continues indefinitely in both directions.

Example 3
Graph $2y = 5x + 10$ using *x*-and *y*-intercepts.

Solution
Find *y*-intercept by setting $x = 0$:
$2y = 5(0) + 10$
$2y = 0 + 10$
$2y = 10$
$y = 5$
Thus, the *y*-intercept is $(0, 5)$.

Find the *x*-intercept by setting $y = 0$.
$2(0) = 5x + 10$

$0 = 5x + 10$

$-10 = 5x$

$\dfrac{-10}{5} = \dfrac{5}{5}x$

$-2 = x$
The *x*-intercept is $(-2, 0)$.

Select a non-zero value of x and find the corresponding value of y and make sure it is collinear with the x- and y-intercepts.

Let $x = 2$. Then $2y = 5(2) + 10$

$$2y = 10 + 10$$

$$2y = 20$$

$$y = 10$$

Thus, $(2, 10)$ is a third point.

Now, plot all 3 points and make sure they are collinear.

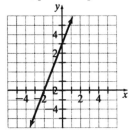

Example 4

Graph $y = -4$.

Solution

Write the equation as $y = 0x - 4$.

Make a table of values:

	$y = 0x - 4$	ordered pair
Let $x = -2$	$y = 0(-2) - 4 = -4$	$(-2, -4)$
Let $x = 0$	$y = 0(0) - 4 = -4$	$(0, -4)$
Let $x = 2$	$y = 0(2) - 4 = -4$	$(2, -4)$

As you can see, no matter what x is, y is always equal to 4.

Plot the ordered pairs.

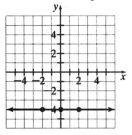

Example 5
Graph the equation $x = 5$.

Solution
Write the equation as $x + 5 = 0y$.
Instead of choosing x-values first, we will choose arbitrary values of y, first.
If $y = -3$, then $x = 5 + 0(-3) = 5$.
So, $(5, -3)$ is a point.
If $y = 0$ then $x = 5 + 0(0) = 5$. So, $(5, 0)$ is a second point.
If $y = 3$ then $x = 5 + 0(3) = 5$. So, $(5, 3)$ is a third point.
As you can see, no matter what value is chosen for y, x is always 5.
Graph.

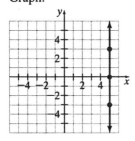

Summary

> 1. The graph of any equation of the form $y = b$ is
>   a horizontal line whose y-intercept is b.
>
> 2. The graph of any equation of the form $x = a$ is
>   a vertical line whose x-intercept is a.

Example 6
The cost of joining a health spa is \$150 plus \$24 per month.

    a. Write an equation for the cost, C, in terms of the number of months, n.

    b. Graph the equation for $0 \le n \le 10$.

    c. Use the graph to estimate the rest for using the spa for 7 months.

Solution

a. To write an equation for the cost, C, in terms of number of months, n, make a table of values:

n	C
0	150
1	$150 + 1(24)$
2	$150 + 2(24)$
10	$150 + 10(24)$

From the pattern of the table, we have $C = 150 + 24n$.

b. The equation is linear. We will find three ordered pairs which solve the equation.

	$C = 150 + 24n$	ordered pair
Let $n = 0$	$C = 150 + 24(0)$	$(0, 150)$
Let $n = 5$	$C = 150 + 24(5)$	$(5, 270)$
Let $n = 10$	$C = 150 + 24(10)$	$(10, 390)$

To graph this equation, we must choose a suitable vertical scale. Notice that when $a = 10$, then $C = 390$. The vertical scale must go from 0 to 400.

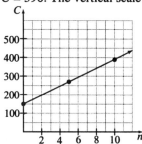

c. Use the graph to estimate the cost for using the spa for 7 months.

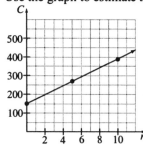

Find the number 7 on the month's axis. Draw a vertical line up to where the graph is intersected. Read the y-value from the graph. It is approximately $320.00. You can obtain the exact value by substituting $n = 7$ into the equation and finding C.
$C = 150 + 24(7) = 150 + 168 = 318$.

Exercise Set 4.2

Find the missing coordinate in the solutions for $3x + y = 9$.

1. $(2, \quad)$ **2.** $(\quad , 3)$ **3.** $(-4, \quad)$

4. $(0, \quad)$ **5.** $(\quad , 0)$

Find the missing coordinate in the solutions for $2x - 7y = 14$.

6. $(0, \quad)$ **7.** $(\quad , 0)$

8. $(-3, \quad)$ **9.** $(\quad , 5)$

Graph each equation

10. $x = -6$ **11.** $y = 5$

12. $y = -3$ **13.** $x = 2$

Graph each equation by plotting points. Plot at least 3 points for each graph.

14. $y = -3x + 2$ **15.** $y = -x + 4$ **16.** $y = 5x + 3$

17. $y = \dfrac{2}{3}x + 1$ **18.** $-2x + 3y = 6$ **19.** $-3x - y = 2$

Graph using the x- and y-intercepts.

20. $4x - y = 8$ **21.** $2x - 3y = 12$

22. $\dfrac{1}{3}x - \dfrac{3}{4}y = 60$ **23.** $\dfrac{2}{3}y = \dfrac{5}{4}x - 120$

24. a. Using the formula $I = prt$, if $p = \$5000.00$, $r = 4\%$, write an equation for simple interest in terms of time, t.

    b. Graph the equations for times of 0 to 20 years inclusive.

    c. What is the simple interest for a time of 8 years?

Answers to Exercise Set 4.2

1. 3   **2.** 2   **3.** 21

4. 9   **5.** 3   **6.** −2

7. 7   **8.** $-\dfrac{20}{7}$ **9.** $\dfrac{49}{2}$

10.
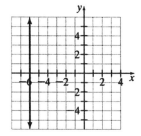

11.
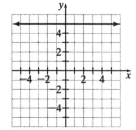

12.
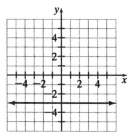

13.
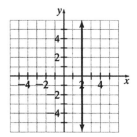

14.
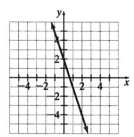

15.
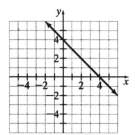

16.
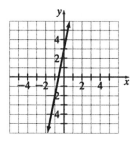

17.
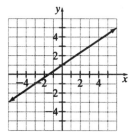

18.
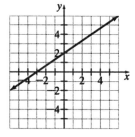

19.
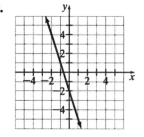

20.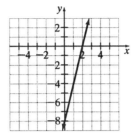

21.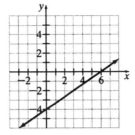

22.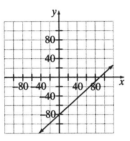

23.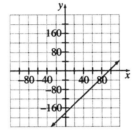

24. a. $I = 5000(0.04)t$

b.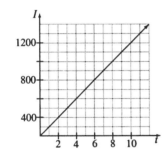

c. $1600

4.3   Slope of a Line

Summary

1. The slope of a line measures the **steepness** of the line.

2. The **slope of a line** is a ratio of the vertical change to the horizontal change between **any two** arbitrary points on the line.

3. The definition of the slope of a line through $(x_1, \ y_1)$ or $(x_2, \ y_2)$ is as follows:

$$\text{Slope} = \frac{y_2 - y_1}{x_2 - x_1}$$

4. It makes no difference which two points are selected when you find the slope. It also makes no difference which point you label as $(x_1, \ y_1)$ or $(x_2, \ y_2)$.

Example 1
Find the slope of the line containing the points (–6, 5) and (3, –4).

Solution

$$\text{Slope} = \frac{y_2 - y_1}{x_2 - x_1} = \frac{5 - (-4)}{-6 - 3} \quad \frac{9}{-9} = -1$$

$$\text{or slope} = \frac{y_1 - y_2}{x_1 - x_2} = \frac{-4 - 5}{3 - (-6)} = \frac{-9}{9} = -1$$

Notice that you can reverse the order of subtraction **provided** you do it for **both the change in** y **as well as the change in** x.

Summary

1. A straight line has **positive** slope if the value of y increases as the value of x increases.

2. A straight line has **negative** slope if the value of y decreases as x increases.

Example 2
Consider the line graphed below:

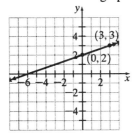

    a. Find the slope of the line using vertical and horizontal changes.

    b. Find the slope using the formula.

Solution

    a. Using the two given points $(0, 2)$ and $(3, 3)$, determine the slope of the line by observing the vertical change and the horizontal change.

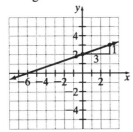

You can see that the slope must be positive since y increases as x increases. The vertical change between the two points is one unit. The horizontal change between the two points is 3 units. Thus, the slope is $\dfrac{1}{3}$.

    b. Find the slope of the line using the two points.

Using $(0, 2)$ and $(3, 3)$ and the slope formula: $m = \dfrac{y_2 - y_1}{x_2 - x_1}$ to obtain $m = \dfrac{3-2}{3-0} = \dfrac{1}{3}$.

Example 3
Find the slope of the line below by observing the vertical change and horizontal change between the two points shown.

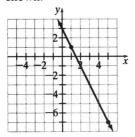

Solution
First, the slope is negative since y decreases as x increases. Next, observe that the vertical change between the two points is –8 and the horizontal change is 4. Thus, the slope, m is $\dfrac{-8}{4} = -2$.

You can verify this by using the slope formula.

$$m = \frac{y_2 - y_1}{x_2 - x_1}.$$

$$m = \frac{-7 - 1}{5 - 1} = \frac{-8}{4} = -2$$

Summary

> 1. Every **horizontal** line has a slope of zero.
>
> 2. The slope of any **vertical** line is undefined.

Example 4
Graph $y = -2$. What is the slope of this line?

Solution
The graph is

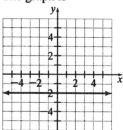

Since this line is horizontal, its slope is zero. Notice that there is **no** change in y between any two points on the line.

Example 5
Graph $x = 5$.

Solution
The graph is

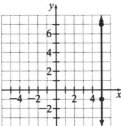

Since this line is vertical, its slope is **undefined**. This line passes through the points $(5, -1)$ and $(5, 7)$. If you calculate the slope using these points, the result is $m = \dfrac{7 - (-1)}{5 - 5} = \dfrac{8}{0}$ (undefined).

Exercise Set 4.3

Find the slope of the line through the given points.

1. $(3, 7)$ and $(6, 16)$ **2.** $(1, 4)$ and $(3, 8)$ **3.** $(3, -4)$ and $(4, -3)$

4. $(0, 5)$ and $(8, -3)$ **5.** $(-7, -5)$ and $(3, -4)$ **6.** $(4, -2)$ and $(-6, 5)$

7. $(5, 7)$ and $(5, -10)$ **8.** $(3, 12)$ and $(-3, 12)$ **9.** $(5, -9)$ and $(-3, -7)$

10. $\left(\dfrac{1}{2}, \dfrac{3}{4} \right), \left(\dfrac{5}{8}, \dfrac{7}{8} \right)$

By observing the vertical and horizontal change of the line between the two points indicated, determine the slope of the line.

11.

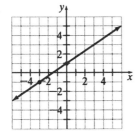

12.

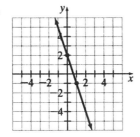

13.

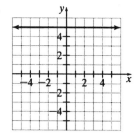

14.

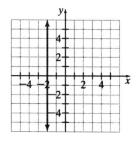

Answers to Exercise Set 4.3

1. 3

2. 2

3. 1

4. −1

5. $\dfrac{1}{10}$

6. $-\dfrac{7}{10}$

7. undefined

8. 0

9. $-\dfrac{1}{4}$

10. 1

11. $m = \dfrac{2}{3}$

12. $m = -3$

13. $m = 0$

14. undefined slope

4.4   Slope-Intercept and Point-Slope Forms of a Linear Equation

Summary

> 1. The **slope-intercept form** of a linear equation is $y = mx + b$ where $m =$ **slope** and $b = y$-**intercept.**
>
> 2. To write a linear equation in **slope-intercept form,** solve the equation for y.

Example 1
Given the following equations of lines, determine the slope and y-intercept.

   a. $y = -2x + 5$ **b.** $y = -\dfrac{3}{4}x - 3$

Solution

   a. $y = -2x + 5$
      Since the coefficient of x is the slope, the slope is -2 and the y-intercept is 5.

   b. $y = -\dfrac{3}{4}x - 3$.

      Here, the coefficient of x is the slope, which is $-\dfrac{3}{4}$, and the y-intercept is -3.

Example 2
Write the equation $2x - 5y = -10$ in slope-intercept form. State the slope and the y-intercept.

Solution
The equation should be solved for y.
$2x - 5y = -10$

$\quad -5y = -2x - 10$

$\quad \dfrac{-5y}{-5} = \dfrac{-2x}{-5} + \dfrac{-10}{-5}$

$\quad\quad y = \dfrac{2}{5}x + 2$

The coefficient of x is the slope, which is $\dfrac{2}{5}$ and the y-intercept is 2.

Summary

> **To Graph Linear Equations Using the Slope and y-intercept:**
>
> 1. Solve the equation for y to obtain $y = mx + b$.
>
> 2. Note the slope, m, and the y-intercept.
>
> 3. Plot the y-intercept on the y-axis.
>
> 4. Use the slope to find a second point on the graph.
>
> 5. Use a straight edge to draw a straight line through the two points.

Example 3
Write the equation $-2x + 5y = -15$ in slope-intercept form. Then, use the slope and y-intercept to graph this line.

Solution
Solve $-2x + 5y = -15$ for y.
$$5y = 2x - 15$$
$$\frac{5y}{5} = \frac{2x}{5} - \frac{15}{5}$$
$$y = \frac{2}{5}x - 3$$

Next, find the slope and y-intercept. The slope is $\frac{2}{5}$ and y-intercept is -3.

Plot the y-intercept $(0, -3)$.
From the slope, we can find a second point on the graph.
From the point $(0, -3)$, move up 2 units and to the right 5 units to obtain a second point $(5, -1)$.
Use a straight-edge to draw a straight line through the two points. A third point can be found by moving up 2 units and to the right 5 units from the point $(5, -1)$. This third point $(10, 1)$ should lie on the straight line.

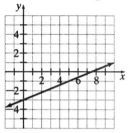

Example 4
Graph $-2x + 3y = 18$ using the slope and y-intercept.

Solution
Solve for y.

$$3y = 2x + 18$$

$$\frac{3y}{3} = \frac{2x}{3} + \frac{18}{3}$$

$$y = \frac{2}{3}x + 6$$

Next, find the slope and y-intercept. Slope is $\frac{2}{3}$ and y-intercept is 6.

Plot the y-intercept $(0, 6)$. From the slope we can find a second point on the graph.
From the point $(0, 6)$, go up 2 units and right 3 units to obtain the point $(3, 8)$. Plot this point.
Use a straight edge to connect the two points. Plot a third point by going up 2 units and right 3 units from the point $(3, 8)$ to obtain $(6, 10)$. Make sure this point lies on the straight line drawn.

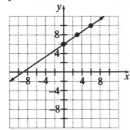

Example 5
Determine the equation of the line using the graph shown.

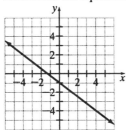

Solution
Since the y-intercept is $(0, -1)$ and the slope is $-\frac{3}{4}$, the equation is $y = -\frac{3}{4}x - 1$.

Summary

> 1. Lines with the **same slope** but **different y-intercepts** are parallel.
>
> 2. If the slopes of the two lines are **not the same,** the lines are **not parallel.**
>
> 3. If two lines have the **same slope** and the **same y-intercept** then both equations represent the **same line.**

Example 6
Determine whether or not the following equations represent parallel lines.
$y = -3x + 1$
$6x - 2y = 7$

Solution
Write each equation in slope-intercept form. The equation $y = -3x + 1$ is already in that form.
Solve $6x - 2y = 7$ for y to obtain
$$-2y = -6x + 7$$

$$\frac{-2y}{-2} = \frac{-6x}{-2} + \frac{7}{-2}$$

$$y = 3x - \frac{7}{2}$$

The slope of the first line is -3 but the slope of the second line is 3. So, the lines are **not parallel.**

Summary

> **Point-slope form of a linear equation:**
>
> $$y - y_1 = m(x - x_1)$$
>
> m is the slope of the line and $(x_1, \ y_1)$ is a point on the line.

Example 7
Write an equation of a line which passes through $(-2, 4)$ and has a slope of 5.

Solution
Use the point-slope form $y - y_1 = m(x - x_1)$ where $(x_1, \ y_1) = (-2, \ 4)$ and has a slope of 5.

$$y - 4 = 5(x - (-2))$$
$$y - 4 = 5(x + 2)$$
$$y - 4 = 5x + 10$$
$$y - 4 + 4 = 5x + 10 + 4$$
$$y = 5x + 14$$

Example 8
Write an equation of the line through the points $(-2, -5)$ and $(3, 10)$.

Solution
First, find the slope of the line.

Use $m = \dfrac{y_2 - y_1}{x_2 - x_1} = \dfrac{10 - (-5)}{3 - (-2)} = \dfrac{15}{5} = 3$.

Now, use the point-slope form of the equation of a line. Use the slope of 3 and either point as $(x_1, \ y_1)$. We will select $(x_1, \ y_1)$ to be $(-2, -5)$.

$$y - y_1 = m(x - x_1)$$
$$y - (-5) = 3(x - (-2))$$
$$y + 5 = 3(x + 2)$$
$$y + 5 = 3x + 6$$
$$y + 5 - 5 = 3x + 6 - 5$$
$$y = 3x + 1$$

Summary

Three methods to graph linear equations
1. Plotting points
2. Using x- and y-intercepts
3. Using the slope and y-intercept

Example 9
Graph $4x - 3y = 9$ by
    a. plotting points.
    b. using intercepts.
    c. using the slope and y-intercept.

Solution

  a. First, solve for y:
$$-3y = -4x + 9$$
$$\frac{-3y}{-3} = \frac{-4x}{-3} + \frac{9}{-3}$$
$$y = \frac{4}{3}x - 3$$

Make a table of ordered pairs.

                                                 Ordered pair:

Let $x = -3$, $y = \frac{4}{3}(-3) - 3 = -7$ $(-3, -7)$

Let $x = 0$, $y = \frac{4}{3}(0) - 3 = -3$ $(0, -3)$

Let $x = 3$, $y = \frac{4}{3}(3) - 3 = 1$ $(3, 1)$

Notice that values of x were chosen that were divisible by 3 because of the fraction $\frac{4}{3}$ in front of the x.

  b. To find the y-intercept, set $x = 0$:
$$4x - 3y = 9$$
$$4(0) - 3y = 9$$
$$-3y = 9$$
$$y = \frac{9}{-3} = -3, \text{ The } y\text{-intercept is } (0, \ -3)$$

To find the x-intercept, set $y = 0$:
$$4x - 3y = 9$$
$$4x - 3(0) = 9$$
$$4x = 9$$
$$x = \frac{9}{4} \text{ or } 2\frac{1}{4}.$$

  c. Put the equation in slope-intercept form by solving it for y.
$$4x - 3y = 9$$
$$-3y = -4x + 9$$
$$y = \frac{-4x}{-3} + \frac{9}{-3}$$
$$= \frac{4}{3}x - 3$$

Then, the slope is $\dfrac{4}{3}$ and the y-intercept is -3.

Using any of these methods, the following graph is obtained.

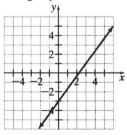

Exercise Set 4.4

Determine the slope and y-intercept of the line represented by each equation. Graph the line using the slope and y-intercept.

1. $y = -2x + 3$ **2.** $3x - y = 5$ **3.** $4x - 4y = 8$

4. $4x - 5y = 20$ **5.** $5x - 4y = 8$

Determine the equation of each line graphed below.

6.

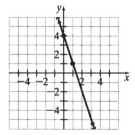

7.

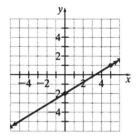

8.

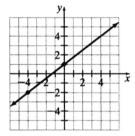

9.

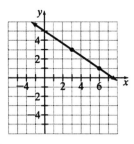

Determine if the lines are parallel.

10. $y = -2x + 5$

$2y + 4x = 0$

11. $3x - 8y = 24$

$y = -\dfrac{3}{7}x + 5$

12. $4x - 7y = 10$

$-8x + 14y = 12$

13. $4x + 2y = 10$

$8x = 4 - 4y$

Write the equation in slope-intercept form.

14. Slope = 5 through (–2, 3)

15. Slope = $\dfrac{3}{2}$ through (–1, 3)

16. Slope = $-\dfrac{1}{3}$ through (3, –4)

17. Through (3, 3) and (7, 7)

18. Through (3, 1) and (6, –1)

19. Through (1, 1) and (5, –15)

20. Graph the equation $-4x + 2y = 6$ by

    a.   plotting points.

    b.   using x- and y-intercepts.

    c.   using slope and y-intercept.

Answers to Exercise Set 4.4

  1. Slope is -2, y-intercept $(0, 3)$

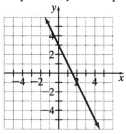

  2. Slope is 3, y-intercept $(0, -5)$

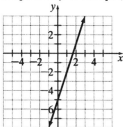

  3. Slope is 1, y-intercept $(0, -2)$

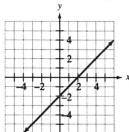

4. Slope is $\dfrac{4}{5}$, *y*-intercept $(0, -4)$

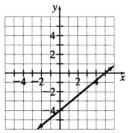

5. Slope is $\dfrac{5}{4}$, *y*-intercept $(0, -2)$

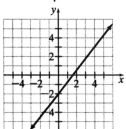

6. $y = -3x + 4$

7. $y = \dfrac{3}{5}x - 2$

8. $y = \dfrac{3}{4}x + 1$

9. $y = -\dfrac{2}{3}x + 5$

10. Yes

11. No

12. Yes

13. Yes

14. $y = 5x + 13$

15. $y = \dfrac{3}{2}x + \dfrac{9}{2}$

16. $y = -\dfrac{1}{3}x - 3$

17. $y = x$

18. $y = -\dfrac{2}{3}x + 3$

19. $y = -4x + 5$

20. a, b, c have same graph

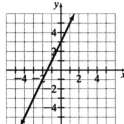

Chapter 4 Practice Test

1. Which of the following ordered pairs satisfy the equation $5x - 10y = 20$?

   a. $(0, 2)$ **b.** $(4, 0)$ **c.** $(0, -2)$

   d. $(-2, 3)$ **e.** $(-2, -3)$

2. Find the slope of the line through the two points $(-4, 5)$ and $(5, 9)$.

3. Write an equation for the graph in the accompanying figure:

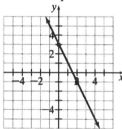

4. Write an equation in slope-intercept form of the line with a slope of -4 passing through $(2\ 5)$.

5. Write in slope-intercept form an equation of the line passing through $(3, 1)$ and $(6, -1)$.

6. Determine if the following equations represent parallel lines.
   Explain your answer.

   $$y = -\frac{3}{4}x + 5$$
   $$3x + 4y = 12$$

7. Find the slope and y-intercept of the line whose equation is $-5x + 3y = 15$.

Graph the following equations.

   8. $x = 7$ 9. $y = -3x + 4$

   10. $-2x + 3y = 9$ 11. $y = -7$

Answers to Chapter 4 Practice Test

   1. b, c, e 2. $m = \dfrac{4}{9}$ 3. $y = -2x + 3$

   4. $y = -4x + 13$ 5. $y = -\dfrac{2}{3}x + 3$

   6. Yes, they are parallel because their slopes are the same. Each slope is equal to $-\dfrac{3}{4}$.

7. Slope is $\dfrac{5}{3}$; *y*-intercept is (0, 5).

8.

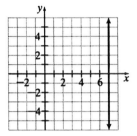

9.

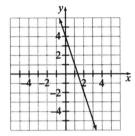

10.

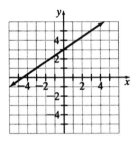

11.

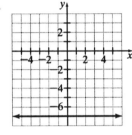

Chapter 5

5.1 Exponents

Summary

In the expression x^n, x is referred to as the **base** and n is called the **exponent**. x^n is read "x to the nth power."

$x^n = x \cdot x \ \dots \ x$
n factors of x

Example 1
Write $x \cdot x \cdot x \cdot y \cdot y$ using exponents

$$\underbrace{x \cdot x \cdot x}_{3 \text{ factors}} \cdot \underbrace{y \cdot y}_{2 \text{ factors}} = x^3 y^2$$

Summary

Product Rule of Exponents

$$x^m \cdot x^n = x^{m+n}$$

Example 2
Multiply using the product rule

Answers

  a. $y^3 \cdot y^2$ $y^3 \cdot y^2 = y^{3+2} = y^5$

Note: We could write $y^3 \cdot y^2$ or $\underbrace{y \cdot y \cdot y}_{3 \text{ factors}} \cdot \underbrace{y \cdot y}_{2 \text{ factors}} = y^5$

  b. $4^2 \cdot 4^7$ $4^2 \cdot 4^7 = 4^{2+7} = 4^9$

  c. $a^5 \cdot a$ $a^5 \cdot a = a^5 \cdot a^1 = a^{5+1} = a^6$

  d. $a^2 \cdot b^4$ $a^2 b^4$ bases are not the same so product rules cannot be applied.

Summary

Quotient Rule
$\dfrac{x^m}{x^n} = x^{m-n}$ $x \neq 0$

Example 3
Divide using the quotient rule

<u>Answers</u>

a. $\dfrac{x^6}{x^4}$

$\dfrac{x^5}{x^4} = x^{6-4} = x^2$

Note: We could write $\dfrac{x^6}{x^4}$ or $\dfrac{x \cdot x \cdot x \cdot x \cdot x \cdot x}{x \cdot x \cdot x \cdot x} = x \cdot x = x^2$

b. $\dfrac{y^8}{y^2}$

$\dfrac{y^8}{y^2} = y^{8-2} = y^6$

c. $\dfrac{7^5}{7^4}$

$\dfrac{7^5}{7^4} = 7^{5-4} = 7^1 = 7$

d. $\dfrac{b^5}{b}$

$\dfrac{b^5}{b} = \dfrac{b^5}{b^1} = b^{5-1} = b^4$

e. $\dfrac{x^3}{y^2}$

$\dfrac{x^3}{y^2}$ bases are not the same so quotient rule cannot be used

Note: $\dfrac{5^4}{5^2} \neq \left(\dfrac{5}{5}\right)^{4-2}$. Do not divide out the bases.

Summary

In this section, to simplify an expression when the numerator and denominator have the same base and the greater exponent is in the denominator, we divide out common factors.

Summary

Zero Exponent Rule

$$x^0 = 1, \; x \neq 0$$

Example 4
Simplify by dividing out a common factor in both the numerator and denominator.

<u>Answers</u>

a. $\dfrac{a^2}{a^5}$

$$\dfrac{a^2}{a^5} = \dfrac{a^2}{a^2 \cdot a^3} = \dfrac{1}{a^3}$$

b. $\dfrac{7^5}{7^6}$

$$\dfrac{7^5}{7^6} = \dfrac{7^5}{7^5 \cdot 7} = \dfrac{1}{7}$$

c. $2a^0$

$$2(a^0) = 2(1) = 2$$

d. $(5a)^0$

$$(5a)^0 = 1$$

Summary

Power Rule for Exponents

$$(x^m)^n = x^{mn}$$

Example 5
Simplify each term.

<u>Answers</u>

a. $(y^2)^4$

$$(y^2)^4 = y^{2 \cdot 4} = y^8$$

b. $(3^3)^5$

$$(3^3)^5 = 3^{3 \cdot 5} = 3^{15}$$

Summary

Expanded Power Rule for Exponents

$$\left(\dfrac{ax}{by}\right)^m = \dfrac{a^m x^m}{b^m y^m} \qquad b \neq 0, y \neq 0$$

125

Example 6
Simplify each expression.

<u>Answers</u>

a. $(3y)^3$

$(3y)^3 = 3^3 y^3 = 27y^3$

b. $(-a)^4$

$(-a^4) = (-1a)^4 = (-1)^4(a^4) = 1 \cdot a^4 = a^4$

c. $(4cd)^2$

$(4cd)^2 = 4^2 c^2 d^2 = 16c^2 d^2$

d. $\left(\dfrac{-4y}{3x}\right)^3$

$\left(\dfrac{-4y}{3x}\right)^3 = \dfrac{(-4)^3 y^3}{3^3 \cdot x^3} = \dfrac{-64y^3}{27x^3}$ or $-\dfrac{64y^3}{27x^3}$

Summary

> Whenever we have an expression raised to a power,
> it helps to simplify the expression in parentheses
> before using the expanded power rule.

Example 7
Simplify each expression.

a. $\left(\dfrac{9x^6 y^3}{3x^3 y^2}\right)^2$ **b.** $(4x^2 y^5)^3 (x^3 y^2)^2$

Solution

a. First, simplify expression within parentheses.
$$\left(\dfrac{9x^6 y^3}{3x^2 y^2}\right)^2 = (3x^4 y)^2$$
Now, apply expanded power rule.
$$(3x^4 y)^2 = 3^2 x^8 y^2 = 9x^8 y^2$$

b. $(4x^2 y^5)^3 (x^3 y^2)^2$
Here, we cannot simplify the expression in parentheses, so use the expanded power rule.
$$(4x^2 y^5)^3 = 4^3 \cdot x^6 y^{15} = 64x^6 y^{15} \text{ and } (x^3 y^2)^2 = x^6 y^4$$
Now apply the product rule.
$$64x^6 y^{15} \cdot x^6 y^4 = 64x^{6+6} \cdot y^{15+4} = 64x^{12} y^{19}$$

Summary of the Rules of Exponents Presented in this Chapter

1. $x^m \cdot x^n = x^{m+n}$ **Product Rule**

2. $\dfrac{x^m}{x^n} = x^{m-n}, \; x \neq 0$ **Quotient Rule**

3. $x^0 = 1, \; x \neq 0$ **Zero Exponent Rule**

4. $(x^m)^n = x^{mn}$ **Power Rule**

5. $\left(\dfrac{ax}{by}\right)^m = \dfrac{a^m x^m}{b^m y^m}, \; b \neq 0, \; y \neq 0$ **Expanded Power Rule**

Exercise Set 5.1

Simplify the following.

1. $c^4 \cdot c^2$

2. $5^3 \cdot 5^5$

3. $(-x)^2 (-x)^3$

4. $r^9 \cdot r^5$

5. $\dfrac{u^6}{u^2}$

6. $\dfrac{n^3}{n^2}$

7. $\dfrac{8^7}{8}$

8. $\dfrac{z^{12}}{z^3}$

9. $(3^5)^2$

10. $(y^3)^7$

11. $(4d^3)^3$

12. $(-3k)^3$

13. $(-1.5)^0$

14. $(6x^5)^0$

15. $\dfrac{x^3 y^8}{x^5 y^5}$

16. $\left(\dfrac{2x^3}{y^2}\right)^2$

17. $\left(\dfrac{a^2 b^3}{a^5 b}\right)^4$

18. $(xy)^2 (2x^2 y^2)^4$

19. $(2a^3 b)^4 (a^2 b^3)$

20. $\left(\dfrac{20x^3 y^7}{5xy^9}\right)^2$

Answers to Exercise Set 5.1

1. c^6

2. 5^{15}

3. $-x^5$

4. r^{14}

5. u^4

6. n

7. 8^6

8. z^9

9. 3^{10}

10. y^{21}

11. $64d^9$

12. $-27k^3$

13. 1

14. 1

15. $\dfrac{y^3}{x^2}$

16. $\dfrac{4x^6}{y^4}$

17. $\dfrac{n^8}{a^{12}}$

18. $16x^{10}y^{10}$

19. $16a^{14}b^7$

20. $\dfrac{16x^4}{y^4}$

5.2   Negative Exponents

Summary

> **Negative Exponent Rule**
>
> $$x^{-m} = \frac{1}{x^m}, \quad x \neq 0$$

Example 1
Use the negative exponent rule to write each expression with positive exponents.

<u>Answers</u>

a. y^{-5} $y^{-5} = \dfrac{1}{y^5}$

b. 3^{-2} $3^{-2} = \dfrac{1}{3^2}$

Note: Raising a number to a negative exponent does not automatically make the value of the expression negative.

c. $\dfrac{1}{a^{-5}}$ $\dfrac{1}{a^{-5}} = \dfrac{1}{\frac{1}{a^5}} = \dfrac{1}{1} \cdot \dfrac{a^5}{1} = a^5$

d. $\dfrac{1}{6^{-1}}$

$$\frac{1}{6^{-1}} = \frac{1}{\frac{1}{6}} = \frac{1}{1} \cdot \frac{6^1}{1} = 6$$

Here are more examples that involve some of preceding rules of exponents along with the negative exponent rule.

Example 2
Simplify.

   a. $(a^3)^{-4}$

   b. $(7^{-3})^2$

Solution

   a. $(a^3)^{-4} = a^{-12}$ Product rule

         $= \dfrac{1}{a^{12}}$ Negative exponent rule

   b. $(7^{-3})^2 = 7^{-6}$ Product rule

         $= \dfrac{1}{7^6}$ Negative exponent rule

Example 3
Simplify and write without negative exponents.

   a. $b^{-5} \cdot b^2$ **b.** $z^{-4} \cdot z^{-3}$ **c.** $\dfrac{n^{-6}}{n^{-2}}$

   d. $\dfrac{12x^2 y^{-3}}{4x^{-1} y^3}$ **e.** $(2x^{-3})^{-4}$

Solution

   a. $b^{-5} \cdot b^2 = b^{-5+2}$ Product rule

         $= b^{-3}$

         $= \dfrac{1}{b^3}$ Negative exponent rule

   b. $z^{-4} \cdot z^{-3} = z^{-4+(-3)}$

         $= z^{-7}$

         $= \dfrac{1}{z^7}$ Negative exponent rule

c. $\dfrac{n^{-6}}{n^{-2}} = n^{-6-(-2)}$ Quotient rule

$= n^{-4}$

$= \dfrac{1}{n^4}$ Negative exponent rule

d. $\dfrac{12x^2y^{-3}}{4x^{-1}y^3} = \dfrac{12}{4} \cdot \dfrac{x^2}{x^{-1}} \cdot \dfrac{y^{-3}}{y^3}$

$= 3x^3y^{-6}$

$= \dfrac{3x^3}{y^6}$ Negative exponent rule

e. $(2x^{-3})^{-4} = 2^{-4}x^{12}$ Expanded power rule

$= \dfrac{x^{12}}{2^4}$ Negative exponent rule

$= \dfrac{x^{12}}{16}$ Simplify

Summary of Rules of Exponents

1. $x^m \cdot x^n = x^{m+n}$ Product Rule

2. $\dfrac{x^m}{x^n} = x^{m-n}, \ x \neq 0$ Quotient Rule

3. $x^0 = 1, \ x \neq 0$ Zero Exponent Rule

4. $(x^m)^n = x^{mn}$ Power rule

5. $\left(\dfrac{ax}{by}\right)^m = \dfrac{a^m x^m}{b^m y^m}, \ b \neq 0, \ y \neq 0$ Expanded Power Rule

6. $x^{-m} = \dfrac{1}{x^m}, \ x \neq 0$ Negative Exponent Rule

Exercise Set 5.2

Simplify each of the following. Write results without negative exponents.

1. a^{-3}

2. $\dfrac{1}{4^{-3}}$

3. $(3x)^{-4}$

4. $(5a)^{-1}$

5. $y^{-2} \cdot y^{-4}$

6. $r^3 \cdot r^{-7}$

7. $\dfrac{k^6}{k^8}$

8. $\dfrac{u^{-7}}{u^{-5}}$

9. $(2x^{-1}y^3)(5xy^{-5})$

10. $\dfrac{18a^{-3}b^5}{6a^2b^{-8}}$

11. $(4c^{-3}d^2)^{-2}$

12. $(2x^5y^{-2})^{-4}$

Answers to Exercise Set 5.2

1. $\dfrac{1}{a^3}$

2. 4^3

3. $\dfrac{1}{81x^4}$ or $\dfrac{1}{3^4 x^4}$

4. $\dfrac{1}{5a}$

5. $\dfrac{1}{y^6}$

6. $\dfrac{1}{r^4}$

7. $\dfrac{1}{k^2}$

8. $\dfrac{1}{u^2}$

9. $\dfrac{10}{y^2}$

10. $\dfrac{3b^{13}}{a^5}$

11. $\dfrac{c^6}{16d^4}$

12. $\dfrac{y^8}{16x^{20}}$

5.3   Scientific notation

Summary

To Write a Number in Scientific Notation

1. Move the decimal point to the right of the first nonzero digit. This will give a number greater than or equal to 1 and less than 10.

2. Count the number of places you moved the decimal to obtain the number in step 1. If the original number was 10 or greater, the count is considered positive. If the original number was less than 1, the count is to be considered negative.

3. Multiply the number obtained in step 1 by 10 raised to the count (power) found in step 2.

Example 1
Write the following numbers in scientific notation

<u>Answers</u>

a. 200, 200

$200, 200 = 2.002 \times 10^5$
Note: We moved the decimal point 5 places and 200,200 is greater than 10, so the power of 10 is positive 5.

b. 0.0056

$0.0056 = 5.6 \times 10^{-3}$
Note: We moved the decimal point 3 places and 0.0056 is less than 10, so the power of 10 is negative 3.

c. 40,000,000

$40,000,000 = 4 \times 10^7$

Summary

> **To convert a Number from Scientific Notation to Decimal Form:**
>
> 1. Obtain the exponent of the power of 10.
>
> 2. **a.** If the exponent is positive, move the decimal in the number (greater than or equal to 1 and less than 10) to the right the same number of places as exponent. It may be necessary to add zeros to the number. This will result in a number greater than or equal to 10.
>
> 2. **b.** If the exponent is 0, do not move the decimal point. Drop the factor 10^0 since it equals 1. This will result in a number greater than or equal to 1.
>
> 2. **c.** If the exponent is negative, move the decimal point in the number to the left the same number of places as the exponent (dropping the negative sign). It may be necessary to add zeros. This will result in a number less than 1.

Example 2
Write each number without exponents.

<u>Answers</u>

a. 5.53×10^3

Moving decimal point 3 places to right gives
$5.53 \times 10^3 = 5.53 \times 1000 = 5530$

b. 6.25×10^{-6}

Moving decimal point 6 places to left gives
$6.25 \times 10^{-6} = 0.00000625$

c. 3.5×10^0

Since exponent is 0, drop the factor 10^0.
So, $3.5 \times 10^0 = 3.5$

We can use the rule of exponents presented in Sections 4.1 and 4.2 when working with numbers written in scientific notation.

Example 3
Multiply $(2.5 \times 10^8)(1.2 \times 10^{-4})$

Solution

$$(2.5 \times 10^8)(1.2 \times 10^{-4}) = (2.5 \times 1.2)(10^8 \times 10^{-4})$$
$$= 3 \times 10^{8-4}$$
$$= 3 \times 10^4 \text{ or } 30,000$$

Example 4

Divide $\dfrac{6 \times 10^{-8}}{2.5 \times 10^{-11}}$

Solution

$$\frac{6 \times 10^{-8}}{2.5 \times 10^{-11}} = \left(\frac{6}{2.5}\right) \times \left(\frac{10^{-8}}{10^{-11}}\right)$$
$$= 2.4 \times 10^{-8-(-11)}$$
$$= 2.4 \times 10^3 \text{ or } 2,400$$

Example 5

Perform the indicated operation by first converting each number to scientific notation. Write the answer in scientific notation

Solution

$$\frac{(2,000,000)(13,000)}{0.00065} = \frac{(2 \times 10^6)(1.3 \times 10^4)}{6.5 \times 10^{-4}}$$
$$= \frac{2 \times 1.3 \times 10^6 \times 10^4}{6.5 \times 10^{-4}}$$
$$= \frac{2.6 \times 10^6 \times 10^4}{6.5 \times 10^{-4}}$$
$$= \left(\frac{2.6}{6.5}\right) \times \left(\frac{10^{10}}{10^{-4}}\right)$$
$$= 0.4 \times 10^{10-(-4)}$$
$$= 0.4 \times 10^{14}$$
$$= 4 \times 10^{-1} \times 10^{14}$$
$$= 4 \times 10^{13}$$

Example 6

Use scientific notation to determine the answer to the following. Write the answer without exponents.

A light year is approximately 5.86 trillion miles (5,860,000,000,000 miles). The diameter of the Milky Way Galaxy is approximately 150,000 light years. What is the diameter of the Milky Way Galaxy in miles?

Solution

$$(5,860,000,000,000)(150,000) = (5.86 \times 10^{12})(1.5 \times 10^5)$$
$$= 6.85 \times 1.5 \times 10^{12} \times 10^5$$
$$= 8.79 \times 10^{12} \times 10^5$$
$$= 8.79 \times 10^{12+5}$$
$$= 8.79 \times 10^{17}$$
$$= 879,000,000,000,000,000$$

This number is read "879 quadrillion."

Exercise Set 5.3

Express each number in scientific notation.

1. 800,000

2. 4,000,000

3. 0.00000075

4. 60

5. 0.000213

6. 3

Express each number without exponents.

7. 4.25×10^7

8. 3.1×10^{-6}

9. 8×10^4

10. 3.14×10^6

11. 4.653×10^0

12. 2.718×10^{-2}

Perform the indicated operation and express each number without exponents.

13. $(2.5 \times 10^5)(1.4 \times 10^4)$

14. $\dfrac{8 \times 10^7}{4 \times 10^3}$

15. $(5 \times 10^6)(8 \times 10^{-10})$

16. $\dfrac{3 \times 10^{-2}}{4 \times 10^{-8}}$

17. $(2.5 \times 10^8)(4 \times 10^{-9})$

Perform the indicated operation by first converting each number to scientific notation. Write the answer in scientific notation.

18. $\dfrac{(0.0072)(0.00003)}{40,000}$

19. $\dfrac{(21,000)(400,000)}{150}$

20. A rocket travels at the rate of 25,000 miles per hour. The distance to Neptune is 2,000,000,000 miles. How many hours will it take the rocket to reach Neptune?

Answers to Exercise Set 5.3

1. 8×10^5

2. 4×10^6

3. 7.5×10^{-7}

4. 6×10^1

5. 2.13×10^{-4}

6. 3×10^0

7. 42,500,000	**8.** 0.0000031	**9.** 80,000
10. 3,140,000	**11.** 4.653	**12.** 0.02718
13. 3,500,000,000	**14.** 20,000	**15.** 0.004
16. 750,000	**17.** 1	**18.** 5.4×10^{-13}
19. 5.6×10^7	**20.** 80,000 hours	

5.4   Addition and Subtraction of Polynomials

Summary

> A **polynomial in x** is an expression containing the sum of a finite number of terms of the form ax^n, for any real number a and any whole number n.

Examples of Polynomials

$4x$

$\dfrac{2}{3}x^2 - 5$

$x^3 - \dfrac{1}{2}x + 4$

Not Polynomials

$3x^{1/3}$
(fractional exponent)

$x^3 + x^{-2}$

(negative exponent)

$\dfrac{4}{x} + x^3$ $\dfrac{4}{x} = 4x^{-1}$
(negative exponent)

Summary

> A polynomial is written in **descending order** or descending **power of the variable** when the exponent on the variable decreases from left to right.

An example of a polynomial in descending order is
$2x^3 + 6x^2 - 8x + 5$

Summary

A **monomial** is a one-term polynomial.

A **binomial** is a two-termed polynomial.

A **trinomial** is a three-termed polynomial.

Summary

The **degree of a term** of a polynomial in one variable is the exponent on the variable of that term.

Examples:

Term	Degree of Term
$6x^3$	third
$3a^7$	seventh
$-2x$	first $(-2x = -2x^1)$
5	zero $(5 = 5x^0)$

Summary

The **degree of a polynomial** in one variable is the same as that of its highest-degree term.

Examples:

Polynomial	Degree of the Polynomial
$4x^5 - 6x - 3$	fifth $(4x^5$ is the highest-degree term$)$
$8x + 1$	first $(8x$ is highest-degree term$)$
15	zero $(15 = 15^0)$

Recall that like terms differ only in their numerical coefficients.

Examples of Like Terms:

6   −3

$7x$ $3x$

$-3y^2$ $6y^2$

$2ab^2$ $-5ab^2$

Summary

> **To add polynomials,** combine the like terms of the polynomials.

Example 1
Simplify the following polynomials:

   a. $(-2x^2 + 5x - 5) + (6x^2 - 8x + 3)$

   b. $2x^3 + 6x - 2) + (x^2 - 3x + 7)$

Solution

   a. $(-2x^2 + 5x - 5) + (6x^2 - 8x + 3)$
      $-2x^2 + 5x - 5 + 6x^2 - 8x + 3$ Remove parentheses
      $-2x^2 + 6x^2 + 5x - 8x - 5 - 3$ Rearrange terms
      $4x^2 - 3x - 8$ Combine like terms

   b. $(2x^3 + 6x - 2) + (x^2 - 3x + 7)$
      $2x^3 + 6x - 2 + x^2 - 3x + 7$ Remove parentheses
      $2x^3 + x^2 + 6x - 3x - 2 + 7$ Rearrange terms
      $2x^3 + x^2 + 3x + 5$ Combine like terms

Summary

To Add Polynomials in Columns

1. Arrange polynomials in descending order one under the other with like terms in the same column.

2. Add the terms in each column.

Example 2
Add the following polynomials using columns.

    a. $-2x^2 + 7x - 6$ and $3x^2 - 3x + 9$

    b. $3y^3 + 6y^2 + 3$ and $3y^2 - 7y - 6$

Solution

    a. $-2x^2 + 7x - 6$ and $3x^2 - 3x + 9$ in columns becomes

$$\begin{array}{r} -2x^2 + 7x - 6 \\ \underline{3x^2 - 3x + 9} \\ x^2 + 4x + 3 \end{array}$$

    b. $3y^3 + 6y^2 + 3$ and $3y^2 - 7y - 6$

$$\begin{array}{r} 3y^3 + 6y^2 + 0y + 3 \\ \underline{3y^2 - 7y - 6} \\ 3y^3 + 9y^2 - 7y - 3 \end{array}$$

Summary

To Subtract Polynomials

1. Remove parentheses. (This will have the effect of changing the sign of every term within the parentheses of the polynomial being subtracted.)

2. Combine like terms.

Example 3
Simplify the following polynomials.

    a. $(6a^2 + 3a - 10) - (2a^2 - 5a - 7)$

    b. $(-x^3 + 3x - 6) - (x^2 - 6x + 2)$

Solution

    a. $(6a^2 + 3a - 10) - (2a^2 - 5a - 7)$

      $6a^2 + 3a - 10 - 2a^2 + 5a + 7$ Remove parentheses (change

                     the sign of each term being subtracted)

      $6a^2 - 2a^2 + 3a + 5a - 10 + 7$ Rearrange terms

      $4a^2 + 8a - 3$ Combine like terms

Note: A common student error is to fail to change the sign of *every* term in the polynomial being subtracted.

    b. $(-x^3 + 3x - 6) - (x^2 - 6x + 2)$

      $-x^3 + 3x - 6 - x^2 + 6x - 2$ Remove parentheses

      $-x^3 - x^2 + 3x + 6x - 6 - 2$ Rearrange terms

      $-x^3 - x^2 + 9x - 8$ Combine like terms

Summary

To Subtract Polynomials in Columns

1. Write the polynomial being subtracted below the polynomial from which it is being subtracted. List like terms in the same column.

2. **Change the sign of each term** in the polynomial being subtracted. (This step can be done mentally if you like.)

3. Add the terms in each column.

Example 4
Subtract using columns.

    a. $(3x^2 - 7x + 11) - (10x^2 + 8x - 5)$

    b. $(5x^3 + 6x - 1) - (3x^2 - 8x - 2)$

Solution

a. $(3x^2 - 7x + 11) - (10x^2 + 8x - 5)$

$$3x^2 - 7x + 11$$
$$\underline{-(10x^2 + 8x - 5)} \quad \text{Align like terms}$$

$$3x^2 - \quad 7x + 11$$
$$\underline{-10x^2 - \quad 8x + 5} \quad \text{Change all signs}$$
$$-7x^2 - 15x + 16 \quad \text{Add}$$

b. $(5x^3 + 6x - 1) - (3x^2 - 8x - 2)$

$$5x^3 + 0x^2 + \ 6x - \ 1 \quad \text{Note } 0x^2 \text{ for missing term}$$
$$\underline{-(3x^2 - \ 8x - 2)} \quad \text{Align like terms}$$

$$5x^3 + 0x^2 + \ 6x - 1$$
$$\underline{-3x^2 + \ 8x + 2} \quad \text{Change all signs}$$
$$5x^3 - 3x^2 + 14x + 1 \quad \text{Add}$$

Exercise Set 5.4

Indicate the expressions that are polynomials. If the polynomials have a specific name—for example monomial or binomial—give that name.

1. $4x^3 + 6x$ **2.** $4x^{1/3 - 6x}$

3. $\dfrac{7}{x} - 1$ **4.** 15

Express each polynomial in descending order. If the polynomial is already in descending order, so state. Give the degree of each polynomial.

5. $2b + b^2 - 8$ **6.** $5x$

7. $-2x^3 + 1 - x^4$ **8.** -3

Add as indicated.

9. $(3u^2 - 7u + 3) + (2u^2 + 2u - 6)$

10. $(6a^2 + 3a - 2) + (4a^2 - 3a - 6)$

11. $(6x^2 - 7x + 1) + (2x^2 - 8)$

12. $(4x^4 - 1) + (4x^4 + 2)$

13. $(-b + 2b^2 - 1) + (b^2 - 6b + 2)$

14. Add in columns. $(3x^2y - 6xy + 5y^2)$ and $(-5x^2y + 8xy - 8y^2)$

Subtract as indicated.

15. $(5x^2 - 3x - 7) - (8x^2 + 2x - 4)$

16. $(-5a^2 + 10a) - (-8a^2 - 11a)$

17. $(2k^2 + 6k + 3) - (5k^2 + 6k + 2)$

18. $(k^3 - k) - (k^2 - 2k - 1)$

19. Subtract in columns. $(10y^3 - 7y^2 - 2y) - (8y^3 + 6y^2 - 3y)$

20. Subtract in columns. $(6y^3 + 10y - 1) - (8y^2 + 3 - 11y)$

Answers to Exercise Set 5.4

1. binomial	**2.** not a polynomial	**3.** not a polynomial
4. not a polynomial	**5.** $b^2 + 2b - 8$, second degree	**6.** $5x$, first degree
7. $-x^4 - 2x^3 + 1$, fourth degree	**8.** -3, zero degree	**9.** $5u^2 - 5u - 3$
10. $10a^2 - 8$	**11.** $8x^2 - 7x - 7$	**12.** $8x^4 + 1$
13. $3b^2 - 7b + 1$	**14.** $-2x^2y + 2xy - 3y^2$	**15.** $-3x^2 - 5x - 3$
16. $3a^2 + 21a$	**17.** $-3k^2 + 1$	**18.** $k^3 - k^2 + k + 1$
19. $2y^3 - 13y^2 + y$	**20.** $6y^3 - 8y^2 + 21y - 4$	

5.5   Multiplication of Polynomials

Summary

> To multiply a monomial by a monomial, multiply their coefficients and use the product rule of exponents to determine the exponent or variable.

Example 1
Multiply

   a. $(5x^4)(3x^3)$

   b. $(-2x^2y^3)(-3x^4y)$

Solution

   a. $(5x^4)(3x^3) = (5 \cdot 3)(x^4 \cdot x^3)$

$$= 15 \cdot x^{4+3}$$

$$= 15x^7$$

   b. $(-2x^2y^3)(-3x^4y) = (-2)(-3)(x^2 \cdot x^4 \cdot y^3 \cdot y^1)$

$$= 6 \cdot x^{2+4} \cdot y^{3+1}$$

$$= 6 \cdot x^6 y^4$$

$$= 6x^6 y^4$$

Summary

> To multiply a polynomial by a monomial, we use the distribution property.
>
> $a(b + c) = ab + ac$
>
> $a(b + c + d + ...n) = ab + ac + ad + ...an$

Example 2
Multiply

   a. $3a(2a - 5)$

   b. $-2c(c^2 - 4c + 2)$

   c. $5y^3(y^2 - 3y - 1)$

Solution

   a. $3a(2a - 5) = (3a)(2a) + (3a)(-5)$

$$= 6a^2 - 15$$

   b. $-2c(c^2 - 4c + 2) = (-2c)(c^2) + (-2c)(-4c) + (-2c)(2)$

$$= -2c^3 + 8c^2 - 4c$$

c. $5y^3(y^2 - 3y - 1) = (5y^3)(y^2) + (5y^3)(-3y) + (5y^3)(-1)$

$$= 5y^5 - 15y^4 - 5y^3$$

Summary

> To multiply a binomial by a binomial use the distributive property in the following manner:
>
> $$(a + b)(c + d) = (a + b)c + (a + b)d$$
> $$= ac + bc + ad + bd$$

Example 3
Multiply $(2x - 3)(x + 4)$

Solution
$(2x - 3)(x + 4) = (2x - 3)(x) + (2x - 3)(4)$

$$= (2x)(x) + (-3)(x) + (2x)(4) + (-3)(4)$$

$$= 2x^2 - 3x + 8x - 12$$

$$= 2x^2 + 5x - 12$$

Summary

> **FOIL**
>
> Consider $(a + b)(c + d)$
>
> **F** stands for first
> $(\underline{a + b})(\underline{c} + d)$ product ac
>
> **O** stands for outer
> $(\underline{a} + b)(c + \underline{d})$ product ad
>
> **I** stands for inner
> $(a + \underline{b})(\underline{c} + d)$ product bc
>
> **L** stands for lasts
> $(a + \underline{b})(c + \underline{d})$ product bd
>
> The product of the two binomials is the sum of these four products.
>
> $\quad\quad$ F $\quad$ O $\quad$ I $\quad$ L
>
> $(a + b)(c + d) = ac + ad + bc + bd$

Example 4
Use the FOIL method to multiply:

    a. $(3x + 1)(2x - 3)$ **b.** $(2x - 3)(2x + 3)$

Solution

    a. $(3x + 1)(2x - 3)$

$$\begin{array}{cccc} \text{F} & \text{O} & \text{I} & \text{L} \end{array}$$
$$= (3x)(2x) + (3x)(-3) + (1)(2x) + (1)(-3)$$
$$= \quad 6x^2 \quad - \quad 9x \quad + \quad 2x \quad - \quad 3$$
$$= \quad 6x^2 - 7x - 3$$

    b. $(2x - 3)(2x + 3)$

$$\begin{array}{cccc} \text{F} & \text{O} & \text{I} & \text{L} \end{array}$$
$$= (2x)(2x) + (2x)(3) + (-3)(2x) + (-3)(3)$$
$$= \quad 4x^2 \quad + \quad 6x \quad - \quad 6x \quad - \quad 9$$
$$= 4x^2 - 9$$

Summary

Product of the Sum and Differences of the Same two Terms

$$(a + b)(a - b) = a^2 - b^2$$

Example 5
Use the rules for finding the product of the sum and difference of two quantities to multiply each expression.

    a. $(x - 5)(x + 5)$

    b. $(3x - 2y)(3x + 2y)$

Solution

    a. $(x - 5)(x + 5)$ use $a = x$ and $b = 5$

$$(a - b)(a + b) = a^2 - b^2$$
$$(x - 5)(x + 5) = (x)^2 - (5)^2$$
$$= x^2 - 25$$

    b. $(3x - 2y)(3x + 2y)$ use $a = 3x$ and $b = 2y$

$$(a - b)(a + b) = a^2 - b^2$$
$$(3x - 2y)(3x + 2y) = (3x)^2 - (2y)^2$$
$$= 9x^2 - 4y^2$$

Summary

> ### Square of Binomial Formulas
>
> $$(a+b)^2 = (a+b)(a+b)$$
> $$= a^2 + 2ab + b^2$$
>
> $$(a-b)^2 = (a-b)(a-b)$$
> $$= a^2 - 2ab + b^2$$

Example 6
Use the square of a binomial formula to multiply each expression.

a. $(x+6)^2$ **b.** $(2t-5c)^2$ **c.** $(k-4)(k-4)$

Solution

a. $(x+6)^2$

$(a+b)^2 = a^2 + 2ab + b^2$ Use $a = x$, $b = 6$

$(x+6)^2 = (x)^2 + 2(x)(6) + (6)^2$

$= x^2 + 12x + 36$

b. $(2t-5c)^2$

$(a-b)^2 = a^2 - 2ab + b^2$ Use $a = 2t$, $b = 5c$

$(2t-5c)^2 = (2t)^2 - 2(2t)(5c) + (5c)^2$

$= 4t^2 - 20tc + 25c^2$

c. $(k-4)(k-4)$

$(k-4)(k-4) = (k-4)^2$

$(a-b)^2 = a^2 - 2ab + b^2$ Use $a = k$, $b = 4$

$(k-4)^2 = k^2 - 2(k)(4) + 4^2$

$= k^2 - 8k + 16$

Note: $(a+b)^2 \neq a^2 + b^2$. A common student error is to forget the middle term $2ab$.

Summary

> When multiplying two polynomials, each term of one polynomial must be multiplied by each term of the other polynomial. We may also multiply a polynomial by a polynomial using a vertical procedure.

146

Example 7

Multiply $(3x + 2)(2x^2 - 5x + 2)$.

Solution

$(3x + 2)(2x^2 - 5x + 2)$

$= 3x(2x^2 - 5x + 2) + 2(2x^2 - 5x + 2)$

$= 6x^3 - 15x^2 + 6x + 4x^2 - 10x + 4$

$= 6x^3 - 11x^2 - 4x + 4$

Using vertical procedure, we have

$$
\begin{array}{r}
2x^2 - 5x + 2 \\
3x + 2 \\
\hline
4x^2 - 10x + 4 \\
6x^3 - 15x^2 + 6x \\
\hline
6x^3 - 11x^2 - 4x + 4
\end{array}
$$

Multiply top polynomial by 2
Multiply top polynomial by $3x$ and align like terms
Add like terms in column

Exercise Set 5.5

Multiply.

1. $-5(3x + 8y)$ **2.** $3y(y^2 + 3y)$ **3.** $-2a(a^4 + 6)$

4. $-2a^2b(a^2 - 3ab + 5b^2)$ **5.** $(x - 1)(x + 5)$ **6.** $(y + 7)(y - 3)$

7. $(4u + 3w)(2u - 3w)$ **8.** $(2c - 7)(3c + 2)$ **9.** $(y^2 + 7)(2y^2 - 5)$

10. $(-d^2 - 1)(-2d^2 - 4)$ **11.** $(2x + 3y)(2x^2 - 3xy + 5y^2)$

12. $(3a - 4b)(9a^2 + 12ab + 16b^2)$

Multiply using a special product formula.

13. $(x - 8)^2$ **14.** $(y + 7)(y - 7)$ **15.** $(x + 3y)^2$

16. $(2a + 5)^2$ **17.** $(r - 6)(r + 6)$ **18.** $(3 - 4a)^2$

19. $(3t - 8u)^2$ **20.** $(4d + 7c)(4d - 7c)$

Answers to Exercise Set 5.5

1. $-15x - 40y$ **2.** $3y^3 + 9y^2$ **3.** $-2a^5 - 12a$

4. $-2a^4b + 6a^3b^2 - 10a^2b^3$ **5.** $x^2 + 4x - 5$ **6.** $y^2 + 4y - 21$

7. $8u^2 - 6uw - 9w^2$ **8.** $6c^2 - 17c - 14$ **9.** $2y^4 + 9y^2 - 35$

10. $2d^4 + 6d^2 + 4$ **11.** $4x^3 + xy^2 + 15y^3$ **12.** $27a^3 - 64b^3$

13. $x^2 - 16x + 64$ **14.** $y^2 - 49$ **15.** $x^2 + 6xy + 9y^2$

16. $4a^2 + 20a + 25$ **17.** $r^2 - 36$ **18.** $16a^2 - 24a + 9$

19. $9t^2 - 48tu + 64u^2$ **20.** $16d^2 - 49c^2$

5.6   Division of Polynomials

Summary

> **To divide a polynomial by a monomial,** divide each term of the polynomial by the monomial.

Example 1
Divide

   a. $\dfrac{8x - 24}{4}$

   b. $\dfrac{12y^4 + 18y^3 - 27y - 9}{37}$

Solution

   a. $\dfrac{8x - 24}{4} = \dfrac{8x}{4} - \dfrac{24}{4}$

   $\qquad\quad = 2x - 6$

   b. $\dfrac{12y^4 + 18y^3 - 27y - 9}{3y} = \dfrac{12y^4}{3y} + \dfrac{18y^3}{3y} - \dfrac{27y}{3y} - \dfrac{9}{3y}$

   $\qquad\qquad\qquad\qquad = 4y^3 + 6y^2 - 9 - \dfrac{3}{y}$

Note: A common student error is to fail to divide each term of the polynomial by the monomial.

$\dfrac{4x + 6}{2} \neq 2x + 6$

Summary

> **Dividing a Polynomial by a Binomial**
>
> When dividing a polynomial by a binomial, use the long division algorithm.

Example 2

Divide $\dfrac{x^2 - 3x - 10}{x + 2}$.

Solution

$\dfrac{x^2 - 3x - 10}{x + 2}$ $\rightarrow$ dividend

$\phantom{\dfrac{x^2 - 3x - 10}{x + 2}}$ $\rightarrow$ divisor

Write as $x + 2 \overline{)x^2 - 3x - 10}$

Divide x^2 (the first term in the dividend) by x (the first term in the divisor).

$\dfrac{x^2}{x} = x$ (quotient)

Place the quotient, x, above the like term containing x in the dividend.

$$x + 2 \overline{)x^2 - 3x - 10} \quad \overset{x}{}$$

Next, multiply x by $x + 2$ as you would do in long division and place the product under their like terms.

$x(x + 2) = x^2 + 2x$

$$\begin{array}{r} x \phantom{{}-3x-10} \\ x + 2 \overline{)x^2 - 3x - 10} \\ x^2 + 2x \end{array}$$

Now, subtract $x^2 + 2x$ from $x^2 - 3x$. Change the signs of the terms being subtracted and add like terms.

$$\begin{array}{r} x \phantom{{}-3x-10} \\ x + 2 \overline{)x^2 - 3x - 10} \\ -x^2 - 2x \\ \hline -5x \end{array}$$

Now, bring down -10, the next term in the dividend.

$$\begin{array}{r} x \phantom{{}-3x-10} \\ x + 2 \overline{)x^2 - 3x - 10} \\ x^2 + 2x \\ \hline -5x - 10 \end{array}$$

Now, divide the first term at the bottom by x, the first term in the divisor.

$\dfrac{-5x}{x} = -5$

Write the -5 in the quotient above the constant in the dividend.

149

$$\begin{array}{r} x-5 \\ x+2\overline{\smash{\big)}\ x^2 - 3x - 10} \\ \underline{x^2 + 2x } \\ -5x - 10 \\ \underline{-5x - 10} \\ 0 \end{array}$$

Subtract (note there is no remainder), so

$$\frac{x^2 - 3x - 10}{x + 2} = x - 5$$

Summary

> **To Check Division of Polynomials**
>
> Divisor $\times$ quotient + remainder = dividend

Example 3

Divide $\dfrac{2x^2 + 3x - 5}{2x - 3}$.

Solution

$$\begin{array}{r} x+3 \\ 2x-3\overline{\smash{\big)}\ 2x^2 + 3x - 5} \\ \underline{2x^2 - 3x } \\ 6x - 5 \\ \underline{6x - 9} \\ 4 \end{array}$$

Thus, $\dfrac{2x^2 + 3x - 5}{2x - 3} = x + 3 + \dfrac{4}{2x - 3}$.

Check: $(2x - 3)(x + 3) + 4 = 2x^2 + 3x - 5$

$2x^2 + 6x - 3x - 9 + 4 = 2x^2 + 3x - 5$

$2x^2 + 3x - 5 = 2x^2 + 3x - 5$ True

Example 4

Divide $18 - 4x + 2x^3$ by $x + 4$.

Solution

First, rewrite the dividend in descending order to get $(2x^3 - 4x + 18) \div (x + 4)$.

Since there is no x^2 term in the dividend, we will add $0x^2$ to help align the terms.

$$
\begin{array}{r}
2x^2 - 8x + 28 \\
x + 4 \overline{\smash{\big)}\, 2x^3 + 0x^2 - 4x + 18} \\
\underline{2x^3 - 8x^2} \\
-8x^2 - 4x \\
\underline{-8x^2 - 32x} \\
28x + 18 \\
\underline{28x + 112} \\
-94
\end{array}
$$

Thus, $\dfrac{2x^3 - 4x + 18}{x + 4} = 2x^2 - 8x + 28 + \dfrac{-94}{x + 4}$.

Summary

Synthetic Division

1. When a polynomial is divided by a binomial of the form $x - a$, the division process can be greatly shortened using a procedure called **synthetic division**.

2. When using synthetic division, the variables are not written since they do not play a role in determining the numerical coefficients of the quotient.

Example 5

Use synthetic division to divide $(4x^3 - 6x^2 + 8x + 2)$ by $(x - 2)$.

Solution

1. Write the dividend in descending powers of x. Then list the numerical coefficients of each term in the dividend. If a term of any degree is missing, place a 0 in the appropriate position to serve as a placeholder.
   4   −6   8   2

2. When dividing by $x - a$, place the "a" to the left of the row of coefficients of the dividend. Our divisor is of the form $x - 2$, so $a = 2$ in this case.
   $2\rfloor$ 4   −6   8   2

3. Bring down the 4. Multiply 4 by 2 (the value of *a*) and place the product 8 underneath the –6.

$$
\begin{array}{r|rrrr}
2 & 4 & -6 & 8 & 2 \\
  &   & 8 &   &   \\
\hline
  & 4 &   &   &   \\
\end{array}
$$

4. Find the sum of –6 and 8 to obtain 2. Multiply 2 by 2 (value of *a*) to obtain 4. Place the 4 under the 8.

$$
\begin{array}{r|rrrr}
2 & 4 & -6 & 8 & 2 \\
  &   & 8 & 4 &   \\
\hline
  & 4 & 2 &   &   \\
\end{array}
$$

5. Find the sum of 8 and 4 and obtain 12. Multiply 12 by 2 (value of *a*) to get 24. Place the 24 underneath the 2. Add 24 and 2 to obtain 26.

$$
\begin{array}{r|rrrr}
2 & 4 & -6 & 8 & 2 \\
  &   & 8 & 4 & 24 \\
\hline
  & 4 & 2 & 12 & 26 \\
\end{array}
$$

6. The first three numbers in the last row, 4, 2, and 12 are the coefficients of the quotient. The last number, 26, is the remainder. The degree of the quotient is **always** one less than the degree of the dividend. Since the dividend is a third-degree polynomial, this means that our quotient is a second-degree polynomial. We can interpret the division problem as follows:

$$(4x^2 - 6x^2 + 8x + 2) \div (x - 2) = (4x^2 + 2x + 12) + \frac{26}{x - 2}$$

Example 6

Find the quotient using synthetic division: $(3x^4 - 25x^2 - 20) \div (x - 3)$

Solution

We can use synthetic division since our divisor is of the form $x - a$. $(a = 3)$

1. Write the coefficients of the dividend in descending powers of *x* and observe that there is a missing x^3 term and a missing *x* term so zeroes will be used for the coefficients of these terms. Place the 3 in front of this row of coefficients.

$$
\begin{array}{r|rrrrr}
3 & 3 & 0 & -25 & 0 & -20 \\
\end{array}
$$

2. Bring down the 3 and multiply 3 by 3 (value of *a*) to obtain 9. Place the 9 underneath 0 and add 0 to 9 to obtain 9.

$$
\begin{array}{r|rrrrr}
3 & 3 & 0 & -25 & 0 & -20 \\
  &   & 9 &   &   &   \\
\hline
  & 3 & 9 &   &   &   \\
\end{array}
$$

3. Repeat the above process to obtain the following result.

$$
\begin{array}{r|rrrrr}
3 & 3 & 0 & -25 & 0 & -20 \\
  &   & 9 & 27 & 6 & 18 \\
\hline
  & 3 & 9 & 2   & 6 & -2
\end{array}
$$

4. Interpret the results. The degree of the dividend is 4. So the degree of the quotient will be 3. The remainder is –2. The remainder will **always be** the last number in the third row.

Thus, $(3x^4 - 25x^2 - 20) \div (x - 3) = 3x^3 + 9x^2 + 2x + 6 + \dfrac{-2}{x-3}$.

Summary

> ### Remainder Theorem
>
> If the polynomial $P(x)$ is divided by $(x - a)$, the remainder is equal to $P(a)$.

Example 7

Use the Remainder Theorem to find the remainder when $8x^3 - 6x^2 - 5x + 3$ is divided by $x + \dfrac{3}{4}$.

Solution

First write the divisor $\left(x + \dfrac{3}{4}\right)$ in the form $x - a$. Since $\left(x + \dfrac{3}{4}\right) = \left(x - \dfrac{-3}{4}\right)$, we will evaluate $P\left(\dfrac{-3}{4}\right)$.

First let $P(x) = 8x^3 - 6x^2 - 5x + 3$.

$$P\left(\frac{-3}{4}\right) = 8\left(\frac{-3}{4}\right)^3 - 6\left(\frac{-3}{4}\right)^2 - 5\left(\frac{-3}{4}\right) + 3$$

$$P\left(\frac{-3}{4}\right) = 8\left(\frac{-27}{64}\right) - 6\left(\frac{9}{16}\right) + \frac{15}{4} + 3$$

$$P\left(\frac{-3}{4}\right) = \frac{-27}{8} - \frac{54}{16} + \frac{15}{4} + \frac{3}{1}$$

$$P\left(\frac{-3}{4}\right) = \frac{-54}{16} + \frac{-54}{16} + \frac{60}{16} + \frac{48}{16} = 0$$

Thus, when $8x^3 - 6x^2 - 5x + 3$ is divided by $\left(x + \dfrac{3}{4}\right)$, the remainder is 0. Since the remainder is 0, this implies that $\left(x + \dfrac{3}{4}\right)$ is a factor of $8x^3 - 6x^2 - 5x + 3$. In general, whenever a polynomial $P(x)$ is divided by $(x - a)$ and the remainder is 0, then $(x - a)$ will be a factor of $P(x)$. To determine the remaining factor, divide $P(x)$ by $x - a$ using synthetic division. The quotient will be the remaining factor.

Example 8

Find the remaining factor of $8x^3 - 6x^2 - 5x + 3$ given that $\left(x + \dfrac{3}{4}\right)$ is a factor of $8x^3 - 6x^2 - 5x + 3$.

Solution

Divide $8x^3 - 6x^2 - 5x + 3$ by $\left(x + \dfrac{3}{4}\right)$ using synthetic division. Recall, $\left(x + \dfrac{3}{4}\right) = \left(x - \dfrac{-3}{4}\right)$. So $a = \dfrac{-3}{4}$.

$$
\begin{array}{r|rrrr}
-\frac{3}{4} & 8 & -6 & -5 & 3 \\
 & & -6 & 9 & -3 \\
\hline
 & 8 & -12 & 4 & 0
\end{array}
$$

Since the dividend was a third-degree polynomial, the quotient in a second-degree polynomial, $8x^2 - 12x + 4$, and is the remaining factor.

Exercise Set 5.6

Divide.

1. $\dfrac{10a - 25}{5}$

2. $\dfrac{6y^2 + 4y}{y}$

3. $\dfrac{3x^2 - 6x}{-3x}$

4. $\dfrac{x^6 - 3x^4 - x^2}{x^2}$

5. $\dfrac{8x^2y^2 - 24xy}{8xy}$

6. $\dfrac{3x^2 - 2x + 1}{x}$

7. $\dfrac{2y^2 - 6y + 9}{y}$

8. $\dfrac{9x^2y + 6xy - 3xy^2}{xy}$

9. $\dfrac{x^2 + 10x + 25}{x + 5}$

10. $\dfrac{y^2 + 2y - 35}{y + 7}$

11. $\dfrac{2y^2 + 7}{y - 3}$

12. $\dfrac{6y^2 + 2y}{2y + 4}$

13. $\dfrac{b^2 - 8b - 9}{b - 3}$

14. $\dfrac{10y^2 + 21y + 10}{2y + 3}$

15. $\dfrac{6a^2 + 25a + 24}{3a - 1}$

16. $\dfrac{4a^3 + 8a^2 + 5a + 9}{2a + 3}$

17. $\dfrac{6a^3 + 5a^2 + 5}{3a - 2}$

18. $\dfrac{4b^3 - b - 5}{2b + 3}$

19. $\dfrac{6x + 2x^3 + 28}{x + 2}$

20. $\dfrac{x^3 - 2x - 4}{x - 2}$

Divide using synthetic division.

21. $(15x^3 - 2x + 7) \div (x + 5)$

22. $(x^3 - 8) \div (x - 2)$

23. $(9x^3 + 9x^2 - x + 2) \div \left(x + \dfrac{2}{3}\right)$

Determine the remainder for the following division problems using the Remainder Theorem. If the divisor is a factor of the dividend, so state.

24. $(x^3 - 6x + 4) \div (x - 1)$ **25.** $(x^4 - 5x^3 - 6x + 30) \div (x - 5)$

Answers to Exercise Set 5.6

1. $2a - 5$

2. $6y + 4$

3. $-x + 2$

4. $x^4 - 3x^2 - 1$

5. $xy - 3$

6. $3x - 2 + \dfrac{1}{x}$

7. $2y - 6 + \dfrac{9}{y}$

8. $9x + 6 - 3y$

9. $x + 5$

10. $y - 5$

11. $2y + 6 + \dfrac{25}{y - 3}$

12. $3y - 5 + \dfrac{20}{2y + 4}$

13. $b - 5 + \dfrac{-24}{b - 3}$

14. $5y + 3 + \dfrac{1}{2y + 3}$

15. $2a + 9 + \dfrac{33}{3a - 1}$

16. $2a^2 + a + 1 + \dfrac{6}{2a + 3}$

17. $2a^2 + 3a + 2 + \dfrac{9}{3a - 2}$

18. $2b^2 - 3b + 4 + \dfrac{-17}{2b + 3}$

19. $2x^2 - 4x + 14$

20. $x^2 + 2x + 2$

21. $15x^2 - 75x + 373 - \dfrac{1858}{x + 5}$

22. $x^2 + 2x + 4$

23. $9x^2 + 3x - 3 + \dfrac{4}{x + \frac{2}{3}}$

24. -1

25. 0; factor

Chapter 5 Practice Test

Simplify each of the following expressions.

1. $4y^3 \cdot 3y^5$

2. $(2x^4)^3$

3. $\dfrac{12a^5}{6a^3}$

4. $\left(\dfrac{2t^5 s^2}{6t^2 s}\right)^3$

5. $(3b^2 c^{-4})^{-3}$

6. $\dfrac{18k^7 v^{-4}}{3k^9 v^2}$

Determine whether each expression is a polynomial. If the polynomial has a specific name, give that name.

7. $x^{-4} + 3x$ **8.** $2x$ **9.** $4x^2 + 3$

10. Write the polynomial $-3 + 4x^5 - 6x^3 - 7x$ in descending order, and give its degree.

Perform the indicated operations.

11. $(2x^2 + 5x - 2) + (6x^2 - 9x + 5)$

12. $(4x^2 - 8x - 5) - (10x^2 + 5x - 11)$

13. $(x^3 + x^2 - 1) - (2x^2 + 5x)$ **14.** $(3x - 1)(2x + 5)$ **15.** $(8 - 3x)(7 + 2x)$

16. $5x^2(2x^2 - 5x - 3)$ **17.** $(3x + 2)(x^2 - 5x + 2)$ **18.** $\dfrac{25y^2 + 15y + 20}{-5}$

19. $\dfrac{8a^4 + 6a^2 + 4}{2a}$ **20.** $\dfrac{8x^2 - 30x + 7}{2x - 7}$

Divide using synthetic division.

21. $(x^2 + x - 14) \div (x - 3)$ **22.** $(5x^2 + 28x - 10) \div (x + 6)$ **23.** $(x^2 + 15x + 56) \div (x - 7)$

Determine the remainder for the following division problems using the Remainder Theorem. If the divisor is a factor of the dividend, so state.

24. $(6x^3 + 5x^2 - 2x + 3) \div (x - 1)$ **25.** $(3y^2 - 13y + 12) \div (y - 3)$

Answers to Chapter 5 Practice Test

1. $12y^8$ **2.** $8x^{12}$ **3.** $2a^2$

4. $\dfrac{t^9 s^3}{27}$ **5.** $\dfrac{c^{12}}{27b^6}$ **6.** $\dfrac{6}{k^2 v^6}$

7. Not a polynomial   **8.** monomial   **9.** binomial

10. $4x^5 - 6x^3 - 7x - 3$, fifth   **11.** $8x^2 - 4x + 3$ **12.** $-6x^2 + 13x + 6$

13. $x^3 - x^2 - 5x - 1$ **14.** $6x^2 + 13x - 5$ **15.** $-6x^2 - 5x + 56$

16. $10x^4 - 25x^3 + 15x^2$ **17.** $3x^3 - 13x^2 - 4x + 4$ **18.** $-5y^2 - 3y - 4$

19. $4a^3 + 3a + \dfrac{2}{a}$ **20.** $4x - 1$ **21.** $x + 4 - \dfrac{2}{x - 3}$

22. $5x - 2 + \dfrac{2}{x+6}$ **23.** $x + 22 + \dfrac{210}{x-7}$ **24.** 12

25. 0; factor

Chapter 6

6.1 Factoring a Monomial from a Polynomial

Summary

1. To factor an expression means to write the expression as a product of its factors.

2. If $a \cdot b = c$, then a and b are factors of c.

3. The greatest common factor (GCF) of two or more numbers is the greatest number that divides all the numbers.

4. To determine the GCF of two or more numbers:

   a. Write each number as a product of prime numbers.

   b. Determine the prime factors common to all the numbers.

   c. Multiply the common factors found in previous step. The product of these factors is the GCF.

Example 1
Find the GCF of 24 and 40.
Write each number as a product of prime numbers.

Solution
$24 = 2 \cdot 2 \cdot 2 \cdot 3$ (Recall 2, 3, and 5 are prime numbers)
$40 = 2 \cdot 2 \cdot 2 \cdot 5$
Factors common to both 24 and 40 are $2 \cdot 2 \cdot 2$
Multiply these factors: $2 \cdot 2 \cdot 2 = 8$
GCF is 8

Example 2
Find the GCF of y^6, y^3, y^5, and y^4.

Solution
The GCF is y^3 since y^3 is the highest power of y that divides each term.

Summary

Greatest Common Factor of two or More Terms

To find the GCF of two or more terms, take each factor the **fewest** number of times it appears in any of the terms.

Example 3
Find the GCF of $a^3b^2c,\ a^2bc^2,\ ab^3c^3$.

Solution
The highest power of a common to all three terms is a^1 or a. The highest power of b common to all three terms is b^1. The highest power of c common to all three terms is c^1 or c. So the GCF of the three terms is abc.

Example 4
Find the GCF of $x^3(x+2)^2,\ x^2(x+2),\ x^2(x+2)^3$.

Solution
The smallest power of x that appears in all three factors is 2 so x^2 is part of the GCF. The smallest power of $(x+2)$ that appears is 1 so $(x+1)$ is part of the GCF. Thus, the GCF is $x^2(x+2)$.

Example 5
Find the GCF of $9x^3,\ 45x^2,\ 21x^5$.

Solution
Since $9 = 3 \cdot 3$, $45 = 3 \cdot 3 \cdot 5$, and $21 = 3 \cdot 7$ the GCF of 9, 45, and 21 is 3. Also the smallest power of x common to all three terms is 2 so x^2 is part of the GCF. Thus, the GCF is $3x^2$.

Summary

To Factor a Monomial from a Polynomial

1. Determine the GCF of all terms in the polynomial.

2. Write each term as the product of the GCF and its other factor.

3. Use the distributive property to factor out the GCF.

4. To check the factoring process, multiply the factors using the distributive property.

Example 6
Factor $8x^2 + 12x$.

Solution
Since $8 = 2 \cdot 2 \cdot 2$ and $12 = 2 \cdot 2 \cdot 3$, then the GCF of 8 and 12 is $2 \cdot 2 = 4$.

The largest power of x that appears is 1 so x^1 is part of the GCF.
Thus, the GCF is $4x$.
Write each term as the product of the GCF and its other factor. $8x^2 + 12x = 4x(2x) + 4x(3)$
Factor out the GCF using the distributive property to get $4x(2x + 3)$.

Example 7
Factor $8y^4 - 12y^2 + 40y$.

Solution
Find GCF: Since $8 = 2 \cdot 2 \cdot 2$, $12 = 2 \cdot 2 \cdot 3$,
$40 = 2 \cdot 2 \cdot 2 \cdot 5$ the GCF of 8, 12, and 40 is 4.
The GCF of y^4, y^2, and y is y.
The GCF of $8y^4 - 12y^2 + 40y$ is $4y$.
$8y^4 - 12y^2 + 40y = 4y(2y^3) + 4y(3y) + 4y(10)$
$$= 4y(2y^3 - 3y + 10)$$
Check: $4y(2y^3 - 3y + 10) = 8y^4 - 12y^2 + 40y$

Example 8
Factor $2x(4x - 3) + 3(4x - 3)$.

Solution

1. The GCF is $(4x - 3)$.

2. Use the distributive property to factor out the GCF: $2x(4x - 3) + 3(4x - 3) = (2x + 3)(4x - 3)$
   The answer may also be expressed as $(4x - 3)(2x + 3)$.

Example 9
Factor $x(x + 4) - (x + 4)$.

Solution
Rewrite: $x(x + 4) - 1(x + 4)$
The GCF is $(x + 4)$.
Factor out the GCF using distributive property
$x(x + 4) - 1(x + 4) = (x - 1)(x + 4)$
or $(x + 4)(x - 1)$

Example 10

Factor $4y^2 + 7xz + 5z^2$

Solution

The only factor common to all 3 terms is one. The polynomial cannot be factored by the method presented in this section.

Summary

> All factoring problems can and should be checked by multiplying the factors. The product of all the factors should equal the original expression. Also, the <u>first</u> step in factoring a polynomial by the methods of this chapter will always be to determine if there is a common factor of each term. You should always factor out the GCF first.

Exercise Set 6.1

Write each number as a product of prime numbers.

1. 70	**2.** 91	**3.** 68
4. 54	**5.** 245	**6.** 175

Find the GCF for the given numbers.

7. 68, 12	**8.** 72, 90	**9.** 51, 34
10. 120, 72		

Find the GCF.

11. $6x,\ 4x^2,\ 8x^3$	**12.** $xy^2,\ x^2y,\ x^2y^2$	**13.** $18x^4,\ 6x^2y,\ 9y^2$
14. $2y - 3, x(2y - 3)$	**15.** $-8x^2,\ -9x^3,\ 12xy^3$	

Factor the GCF from each term in the expression. If an expression cannot be factored, so state.

16. $3x + 3y$	**17.** $15y - 5$	**18.** $9y^2 - 6y$
19. $36a^{12} - 24a^6$	**20.** $y + 3yx^2$	**21.** $7x + 5y$
22. $(m + n)(m + n) - (m + n)$	**23.** $52x^2y^2 + 16xy^3 + 26z$	**24.** $4y(2x + 1) + (2x + 1)$
25. $15x^2 - 16x + 7$		

Answers to Exercise Set 6.1

1. $2 \cdot 5 \cdot 7$

2. $13 \cdot 7$

3. $2^2 \cdot 17$

4. $2 \cdot 3^3$

5. $5 \cdot 7^2$

6. $5^2 \cdot 7$

7. 4

8. 18

9. 17

10. 24

11. $2x$

12. xy

13. 3

14. $2y - 3$

15. x

16. $3(x + y)$

17. $5(3y - 1)$

18. $3y(3y - 2)$

19. $12a^6(3a^6 - 2)$

20. $y(1 + 3x^2)$

21. cannot be factored

22. $(m + n)(m + n - 1)$

23. $2(26x^2y^2 + 8xy^3 + 13z)$

24. $(2x + 1)(4y + 1)$

25. cannot be factored

6.2 Factoring By Grouping

Summary

To Factor a Four-Term Polynomial Using Grouping

1. Determine if there are any factors common to all four terms. If so, factor the GCF from each of the four terms.

2. If necessary, arrange the terms so that the first 2 terms have a common factor and the last two have a common factor.

3. Use the distributive property to factor each group of two terms.

4. Factor the GCF from the result of step 3.

Example 1
Factor by grouping: $x^2 + 5x + 2x + 10$

Solution
Check to see if there are common factors among the four terms. In this case there are none.
Rearrange: $(x^2 + 5x) + (2x + 10)$
$(x^2 + 5x) + (2x + 10) = x(x + 5) + 2(x + 5)$
Factor out the common binomial factor of $x + 5$: $(x + 2)(x + 5)$

Note: The like terms $5x$ and $2x$ of this example could have been combined. But our purpose of this example was to illustrate the method of factoring by grouping, so we did not combine them.

Example 2
Factor $3x^2 - 15x + 4x - 20$ by grouping.

Solution
No common factor among all four terms
Rearrange: $(3x^2 - 15x) + (4x - 20)$

Note: $3x$ is common to first pair and 4 is common to last pair.

$3x(x - 5) + 4(x - 5)$
Factor out common binomial of $(x - 5)$
$(3x + 4)(x - 5)$ or $(x - 5)(3x + 4)$

Example 3
Factor $5x^2 - 25x - x + 5$ by grouping.

Solution
No common factor among all four terms
Rearrange: $(5x^2 - 25x) - x + 5$
$5x$ is GCF of first pair
$5x(x - 5) - x + 5$
Here, we must factor out a negative one from the second binomial to make it identical with the factor $(x - 5)$ of the first pair:
$5x(x - 5) - 1(x - 5)$
Factor out common binomial $(x - 5)$:
$(5x - 1)(x - 5)$

Example 4
Factor $-x^3 + bx^2 + 3b^2x - 3b^3$ by grouping

Solution
No common factor among all four terms

Rearrange: $(-x^3 + bx^2) + (3b^2x - 3b^3)$
x^2 is common to first pair and $3b^2$ is GCF of second pair.
$x^2(-x + b) + 3b^2(x - b)$

Here, we must adjust the sign of one of the binomial factors to make it identical with the other binomial term.

Factor out a negative one from first binomial: $x^2(-1)(x - b) + 3b^2(x - b)$.
$(-x^2 + 3b^2)(x - b)$ or $(3b^2 - x^2)(x - b)$

Example 5

Factor $6x + 5y + xy + 30$.

Solution

No common factor among all four terms

Note: No common factors among the first pair.

Rearrange: $6x + xy + 5y + 30$

$(6x + xy) + (5y + 30)$

x is common to first pair and y is common to second pair.

$6x + xy + 5y + 30 = x(y + 6) + 5(y + 6)$

Factor out common binomial $(y + 6)$:

$(x + 5)(y + 6)$

Example 6

Factor $2x^3 - 5x^2 - 6x^2 + 15x$ by grouping.

Solution

Notice that x is the GCF.

$x(2x^2 - 5x - 6x + 15)$

$x[(2x^2 - 5x) - 6x + 15] = x[x(2x - 5) - 3(2x - 5)]$

$\qquad\qquad\qquad\qquad = x(x - 3)(2x - 5)$

Exercise Set 6.2

Factor by grouping.

1. $x^3 - x^2 + 7x - 7$

2. $ax - ay + by - by$

3. $x^2 + bx - ax - ab$

4. $y^3 + 2y^2 - 4y - 8$

5. $x^2 - 3x + 2xy - 6y$

6. $x^2 - 2b + 3ax - 6ab$

7. $2t^2 - 10t + 3t - 15$

8. $x^3 + 2x^2 + x^2 + 2x$

9. $y^2 - yb + ya - ab$

10. $18m - 12m^2 + 24m^3 - 16m^4$

11. $x^2 - x - x + 1$

12. $y^2 - 3y - 3y + 9$

13. $ax + ay - bx - by$

14. $2y^2 - 4xy + 8xy - 16x^2$

15. $a^3 - 2a^2b + 2ab^2 - 4b^3$

16. $2cx + vx + 4cy + 2vy$

17. $36x^3 - 18x^2 - 84x + 42$

18. $y^2 - 4by + 3ay - 12ab$

19. $xy - x + 5y - 5$

20. $xa - xb + ca - cb$

Answers to Exercise Set 6.2

1. $(x^2 + 7)(x - 1)$

2. $(x - y)(a + b)$

3. $(x + b)(x - a)$

4. $(y^2 - 4)(y + 2)$

5. $(x + 2y)(x - 3)$

6. $(x + 3a)(x - 2b)$

7. $(2t + 3)(t - 5)$

8. $x(x + 2)(x + 1)$

9. $(y + a)(y - b)$

10. $2m(3 + 4m^2)(3 - 2m)$

11. $(x - 1)(x - 1) = (x - 1)^2$

12. $(y - 3)(y - 3) = (y - 3)^2$

13. $(a - b)(x + y)$

14. $(2y + 8x)(y - 2x)$

15. $(a^2 + 2b^2)(a - 2b)$

16. $(2c + v)(x + 2y)$

17. $6(3x^2 - 7)(3x - 1)$

18. $(y - 4b)(y + 3a)$

19. $(x + 5)(y - 1)$

20. $(x + c)(a - b)$

6.3   Factoring Trinomials of the form $ax^2 + bx + c$, $a = 1$

Summary

> **To Factor Trinomials of the form $ax^2 + bx + c$, where $a = 1$:**
>
> 1. Find two numbers whose product equals the constant c and whose sum equals the coefficient b.
>
> 2. Using the two numbers found in step one, the factored form of $ax^2 + bx + c$ will be $(x +$ one number$)(x +$ second number$)$.
>
> 3. Hint: If the constant c is positive, then both factors must have the same sign; either both are positive or both are negative. If the constant c is negative, then the two factors will have opposite signs.

Example 1

Factor $x^2 - 7x + 12$.

Solution
Since c is positive both factors must have the same sign, negative, since the coefficient of x is -7.
Make a chart:

Factors of +12	Associated Sums
$(-1)(20)$	19
$(1)(-20)$	-19
$(2)(-10)$	-8
$(-2)(10)$	8

Thus, -3 and -4 are the required factors.
The factorization is $(x-3)(x-4)$.

Check by FOIL: $(x-3)(x-4) = x^2 - 4x - 3x + 12$
$$= x^2 - 7x + 12$$

Example 2
Factor $x^2 + 8x - 20$.

Solution
Since $c = -20$ is negative, required factors must have opposite signs.

Factors of -20	Associated Sums
$(-1)(20)$	19
$(1)(-20)$	-19
$(2)(-10)$	-8
$(-2)(10)$	8

(We can stop here, since we have found the desired factors)

-2 and 10 are the required numbers.

$x^2 + 8x - 20 = (x + 10)(x - 2)$
Check: $(x + 10)(x - 2) = x^2 - 2x + 10x - 20$
$$= x^2 + 8x - 20$$

Example 3
Factor $y^2 + 11y - 30$.

Solution
Since $c = -30$ is negative, the factors must have opposite signs.

Factors of –30	Associated Sums
(–1)(30)	29
(1)(–30)	–29
(2)(–15)	–13
(–2)(15)	13
(3)(–10)	–7
(5)(–6)	–1
(–5)(6)	1

Since we cannot find any factors of –30 whose sum is 11, we conclude that the trinomial cannot be factored and is prime.

Example 4
Factor $x^2 - 13xy - 48y^2$.

Solution
This example has two variables but the procedure is essentially the same. Just remember that the product of the last terms of the binomials must be $-48y^2$.

Since –48 is negative, the factors must have opposite signs.

Factors of –48	Associated Sums
(–1)(48)	47
(1)(–48)	–47
(2)(–24)	–22
(–2)(24)	22
(3)(–16)	–13

Thus, 3 and –16 are the desired factors.

Write the answer $(x + 3y)(x - 16y)$
Check: $(x + 3y)(x - 16y) = x^2 - 16xy + 3xy - 48y^2$
$$= x^2 - 13xy - 48y^2$$

Example 5

Factor $x^2 - 11x - 60$ using trial and error.

Solution

Since the last number, –60, is negative, the factors will have opposite signs.

Factors of –60	Possible Factors
$(-60)(1)$	$(x-60)(x+1) = x^2 - 59x - 60$
$(-30)(2)$	$(x-30)(x+2) = x^2 - 28x - 60$
$(-20)(3)$	$(x-20)(x+3) = x^2 - 17x - 60$
$(-15)(4)$	$(x-15)(x+4) = x^2 - 11x - 60$

We stop at this point.

So, $x^2 - 11x - 60 = (x-15)(x+4)$.

Example 6

Factor $2x^2 + 8x - 10$ using any method.

Solution

Since $a \ne 1$, we check for a common factor, 2 is the GCF. So we factor the 2 out.
$2(x^2 + 4x - 5)$

Factors of –5	Associated Sums
$(-5)(1)$	–4
$(5)(-1)$	4

$2(x^2 + 4x - 5) = 2(x+5)(x-1)$

Example 7

Factor $2x^3 - 12x^2 + 10x$.

Solution

Factor out the GCF of $2x$:
$2x(x^2 - 6x + 5)$

Factors of 5	Associated Sums
$(1)(5)$	6
$(-1)(-5)$	–6

$2x^3 + 12x^2 + 10x = 2x(x-1)(x-5)$

Exercise Set 6.3

Factor each expression if possible. If not possible, then so state.

1. $y^2 + 3y + 2$ **2.** $y^2 - 2y - 15$ **3.** $b^2 + 4b - 21$

4. $x^2 + 2x - 35$ **5.** $a^2 + 6a + 9$ **6.** $x^2 - 8x + 12$

7. $c^2 + cx - 56x^2$ **8.** $x^2 - 9xy + 20y^2$ **9.** $m^2 + 4m + 6$

10. $-4y^2 + 4y + 24$ **11.** $b^2 - 15bz + 36z^2$ **12.** $4x^2 + 12x - 16$

13. $x^2 - 6xy + 8y^2$ **14.** $x^3 + 11x^2 - 42x$ **15.** $2x^3 - 4x^2 - 30x$

16. $x^2 - 17x - 60$ **17.** $3x^2 - 6x - 24$ **18.** $x^2 - 14x - 51$

19. $x^2 + 21x + 68$ **20.** $x^2 - 26x + 69$ **21.** $x^2 + 7x + 24$

22. $x^2 - 18x - 40$ **23.** $x^2 + 38x + 72$ **24.** $x^2 - 28xy - 60y^2$

25. $2x^3 + 6x^2 - 56x$

Answers to Exercise Set 6.3

1. $(y + 2)(y + 1)$ **2.** $(y - 5)(y + 3)$ **3.** $(b + 7)(b - 3)$

4. $(x + 7)(x - 5)$ **5.** $(a + 3)(a + 3)$ **6.** $(x - 6)(x - 2)$

7. $(c + 8x)(c - 7x)$ **8.** $(x - 4y)(x - 5y)$ **9.** Does not factor

10. $-4(y - 3)(y + 2)$ **11.** $(b - 3z)(b - 12z)$ **12.** $4(x + 4)(x - 1)$

13. $(x - 4y)(x - 2y)$ **14.** $x(x + 14)(x - 3)$ **15.** $2x(x - 5)(x + 3)$

16. $(x - 20)(x + 3)$ **17.** $3(x - 4)(x + 2)$ **18.** $(x - 17)(x + 3)$

19. $(x + 17)(x + 4)$ **20.** $(x - 23)(x - 3)$ **21.** Does not factor

22. $(x - 20)(x + 2)$ **23.** $(x + 36)(x + 2)$ **24.** $(x - 30y)(x + 2y)$

25. $2x(x + 7)(x - 4)$

6.4   Factoring Trinomials of the form $ax^2 + bx + c$, $a \neq 1$

Summary

One method of factoring trinomials of form
$ax^2 + bx + c$, $a \neq 1$, is the trial and error method.

1. Determine if there are any common factors to all three terms. If so, factor them out.

2. Write all pairs of factors of the coefficient of the squared term, a.

3. Write all pairs of factors of the constant term, c.

4. Try combinations of these factors until the correct middle term, bx, is found.

Example 1

Factor $3x^2 + 8x + 4$ using trial and error.

Solution

Factors of 4	Possible Factors of Trinomial	Sum of the products of the outer and inner terms
1(4)	$(3x + 1)(x + 4)$	$12x + x = 13x$
4(1)	$(3x + 4)(x + 1)$	$3x + 4x = 7x$
2(2)	$(3x + 2)(x + 2)$	$6x + 2x = 8x$

Since $(3x + 2)(x + 2)$ yields the correct middle term, they are the correct factors.

Hint: Since the sign of 4, the constant term, is positive then the factors or 4 must have the same sign. But the sign of the coefficient of the middle term, $8x$, is positive, so it was only necessary to consider the positive factors of 4.

Example 2
Factor $2x^2 - 15x - 8$ using trial and error.

Solution

Factors of –8	Possible Factors of Trinomial	Sum of the products of the outer and inner terms
–8(1)	$(2x - 8)(x + 1)$	$-6x$
–4(2)	$(2x - 4)(x + 2)$	$0x$
8(–1)	$(2x + 8)(x - 1)$	$6x$
4(–2)	$(2x + 4)(x - 2)$	$0x$
(–1)(8)	$(2x - 1)(x + 8)$	$15x$
(1)(–8)	$(2x + 1)(x - 8)$	$-15x$

Thus, our desired factors are $(2x + 1)(x - 8)$ since the sum of the products of the outer and inner terms is $-15x$.

Example 3
Factor $6x^3 + 5x^2 - 6x$ by trial and error.

Solution
Remove common x factor: $x(6x^2 + 5x - 6)$

Now, factor $6x^2 + 5y - 6$:

Factors of –6	Possible Factors of Trinomial	Sum of the products of the outer and inner terms
–6(1)	$(2x - 6)(3x + 1)$	$-16x$
–3(2)	$(2x - 3)(3x + 2)$	$-5x$
6(–1)	$(2x + 6)(3x - 1)$	$16x$
3(–2)	$(2x + 3)(3x - 2)$	$5x$

Since $(2x + 3)(3x - 2)$ yields the correct sum of the products of the outer and inner terms, this is the correct factorization.

Thus, $6x^3 + 5x^2 - 6x = x(2x + 3)(3x - 2)$.

Summary

A second method for factoring $ax^2 + bx + c,\ a \neq 1,$ is the method of grouping.

1. Factor out the GCF of the three terms.

2. Find two numbers whose product is equal to ac and whose sum is equal to b.

3. Rewrite the middle term, bx, as the sum or difference of two terms using the numbers found in step 2.

4. Factor by grouping.

Example 4
Factor $3x^2 + 17x + 10$.

Solution
GCF is 1
Let $a = 3, b = 7, c = 10$

Now set up a chart:

Factors of $ac = 30$	Associated Sums
1(30)	$30 + 1 = 31$
2(15)	$2 + 15 = 17$

Since $2 \cdot 15 = 30$ and $2 + 15 = 17$, then 2 and 15 are the desired numbers.

Rewrite $3x^2 + 2x + 15x + 10 = (3x^2 + 2x) + (15 + 10)$.
Factor by grouping: $(3x^2 + 2x) + (15x + 10)$
$\qquad\qquad x(3x + 2) + 5(3x + 2)$
$\qquad\qquad (x + 5)(3x + 2)$
Thus, $3x^2 + 17x + 12 = (x + 5)(3x + 2)$.

Example 5
Factor $5x^2 - 11x - 12$.

Solution
GCF is 1
Let $a = 5, b = -11, c = -12$

Now, set up a chart:

Factors of $ac = -60$	Associated Sums
$-1(60)$	$-1 + 60 = 59$
$-2(30)$	$-2 + 30 = 28$
$-3(20)$	$-3 + 20 = 17$
$-4(15)$	$-4 + 15 = 11$
$4(-15)$	$4 + (-15) = -11$

Thus, 4, -15 are the required numbers.
Rewrite $5x^2 + 4x - 15x - 12$ as $(5x^2 + 4x) - 15x - 12$.
Factor by grouping: $(5x^2 + 4x) - 15x - 12$
$$x(5x + 4) - 3(5x + 4)$$
$$(x - 3)(5x + 4)$$
Notice: In this step, it was necessary to factor out a -3 rather than a 3 from the last two terms.
Thus, $5x^2 - 11x - 12 = (x - 3)(5x + 4)$.

Example 6
Factor $3x^2 - 7x - 8$.

Solution
GCF is 1
Let $a = 3, b = -7, c = -8$

Now, set up a chart:

Factors of $ac = -24$	Associated Sums
$-24(1)$	-23
$-12(2)$	-10
$-8(3)$	-5
$-6(4)$	-2
$-4(6)$	2
$-2(12)$	10

We can see it as not possible to find factors of -24 whose sum is -7 so we conclude this trinomial is not factorable.

Exercise Set 6.4

Factor completely. If an expression cannot be factored, so state.

1. $6x^2 - 11x - 3$ 2. $2x^2 + 5x - 7$ 3. $3x^2 - 5x + 2$

4. $10x^2 + x - 3$ 5. $6a^2 + 5ab - 6b^2$ 6. $6x^2 - 23x + 20$

7. $6x^2 + x - 1$ 8. $4x^2 - 5x + 1$ 9. $12u^2 - 11uv + 2v^2$

10. $42x^2 - 10xy - 6y^2$ 11. $10m^2 - 24m + 8$ 12. $6b^2 - 3b - 84$

13. $5x^2 + 2x + 7$ 14. $8x^2 + 13x - 6$ 15. $15x^2 - 19x + 6$

16. $10x^3 + 17x^2 + 3x$ 17. $9x^2 - 6x + 1$ 18. $25x^2 - 20x + 4$

19. $56y^2 + 5yz - 6z^2$ 20. $2x^2 - 7xy + 3y^2$

Answers to Exercise Set 6.4

Factor completely. If an expression cannot be factored, so state.

1. Cannot be factored   2. $(x-1)(2x+7)$ 3. $(x-1)(3x-2)$

4. $(5x+3)(2x-1)$ 5. $(2a+3b)(3a-2b)$ 6. $(3x-4)(2x-5)$

7. $(2x+1)(3x-1)$ 8. $(x-1)(4x-1)$ 9. $(3u-2v)(4u-v)$

10. $2(3x-2y)(7x+3)$ 11. $2(5m-2)(m-2)$ 12. $3(2b+7)(b-4)$

13. Cannot be factored   14. $(8x-3)(x+2)$ 15. $(5x-3)(3x-2)$

16. $x(5x + 1)(2x + 3)$ **17.** $(3x - 1)^2$ **18.** $(5x - 2)^2$

19. $(8y + 3z)(7y - 2z)$ **20.** $(2x - y)(x - 3y)$

6.5   Special Factoring Formulas and a General Review of Factoring

Summary

```
┌─────────────────────────────────────────────┐
│          Difference of Two Squares           │
│  a² − b² = (a + b)(a − b)                     │
└─────────────────────────────────────────────┘
```

$$a^2 - b^2 = (a + b)(a - b)$$

Example 1
Factor each of the following using the difference of two squares formula.

a. $x^2 - 16 = x^2 - 4^2$
  Use $a^2 - b^2 = (a + b)(a - b)$ with $a = x$ and $b = y$ to get
  $$x^2 - 4^2 = (x + 4)(x - 4)$$

b. $9 - 16x^2 = 3^2 - (4y)^2$
  Use $a^2 - b^2 = (a + b)(a - b)$ with $a = 3$ and $b = 4y$ to get
  $$3^2 - (4y)^2 = (3 + 4y)(3 - 4y)$$

c. $49x^2 - 9y^2 = (7x)^2 - (3y)^2$
  Use $a^2 - b^2 = (a + b)(a - b)$ with $a = 7x$ and $b = 3y$ to get
  $$(7x)^2 - (3y)^2 = (7x + 3y)(7x - 3y)$$

d. $9y^4 - 16x^4 = (3y^2)^2 - (4x^2)^2$
  Use $a^2 - b^2 = (a + b)(a - b)$ with $a = 3y^2$ and $b = 4x^2$ to get
  $$(3y^2)^2 - (4x^2)^2 = (3y^2 + 4x^2)(3y^2 - 4x^2)$$

e. $8x^2 - 72v^2$
  First, remove the GCF of 8.
  $$8(x^2 - 9y^2) = 8[(x)^2 - (3y)^2]: \ a = x, \ b = 3y$$
  $$= 8(x + 3y)(x - 3)$$

f. $x^4 - 1 = (x^2)^2 - 1^2$
  $$= (x^2 - 1)(x^2 + 1): \ a = x^2, \ b = 1$$
  We can factor $x^2 - 1$ as $(x + 1)(x - 1)$ so our final answer is:
  $$x^4 - 1 = (x + 1)(x - 1)(x^2 + 1)$$
  Remember: Always factor <u>completely</u>.

It is helpful to recall numbers which are perfect squares. The first few perfect squares are: 1, 4, 9, 16, 25, 36, 49, 64, 81, 100, 121, 144, 169, 196, 225 and so on.

Also, any variable raised to an even power will be a perfect square.

$$x^2, \ x^4 = (x^2)^2 \qquad\qquad x^6 = (x^3)^2 \qquad\qquad x^8 = (x^4)^2$$

Example 2

Factor $144x^6 - 121y^4$.

Solution

$$144x^6 - 121y^4 = (12x^3)^2 - (11y^2)^2$$
$$= (12x^3 + 11y^2)(12x^3 - 11y^2)$$

Example 3

Factor $x^2 + 9$.

Solution

This is a sum of two perfect squares—not a difference of two perfect squares. It cannot be factored further using the set of real number.

Summary

Sum of two cubes

$$a^3 + b^3 = (a+b)(a^2 - ab + b^2)$$

Difference of two cubes

$$a^3 - b^3 = (a-b)(a^2 + ab + b^2)$$

Example 4

Factor, using the sum of two cubes formula or the difference of two cubes formula.

    a. $x^3 + 8 = x^3 + 2^3$

        Use $a^3 + b^3 = (a+b)(a^2 - ab + b^2)$ with $a = x$ and $b = 2$ to get

           $x^3 + 2^3 = (x+2)(x^2 - 2x + 4)$

    b. $m^3 - 64 = m^3 - 4^3$

        Use $a^3 - b^3 = (a-b)(a^2 + ab + b^2)$ with $a = m$ and $b = 4$ to get

           $m^3 - 4^3 = (m-4)(m^2 + 4m + 16)$

    c. $x^5 - 27x^2$

        First, remove GCF of x^2: $x^2(x^3 - 27)$

Then, factor $x^3 - 27 = x^3 - 3^3$
$$= (x-3)(x^2 + 3x + 9)$$

Final answer is $x^5 - 27x^2 = x^2(x-3)(x^2 + 3x + 9)$.

It is helpful to memorize numbers which are perfect cubes:
$1, 8, 27, 64, 125, 216, 343, 512, 729, 1000, \ldots$
Also, variables which are raised to exponents which are multiples of 3 are perfect cubes:
$$x^3, \ x^6 = (x^2)^3, \qquad\qquad x^9 = (x^3)^3, \qquad\qquad x^{12} = (x^4)^3, \ \ldots$$

Example 5
Factor $8x^6 + 125$.

Solution
$$8x^6 + 125 = (2x^2)^3 + 5^3 = (2x^2 + 5)(4x^4 - 10x^2 + 25)$$

Summary

General Procedure to Factor a Polynomial

1. If all the terms of the polynomial have a GCF other than 1, **FACTOR IT OUT.**

2. If the polynomial has two **terms**, determine if it is a **difference of two squares**, or a **sum or difference of two cubes.** If so, factor it using the appropriate formula.

3. If a polynomial has **three terms**, factor it using either **trial and error** or by **grouping.**

4. If the polynomial has **more than three terms** try **factoring by grouping.**

5. As a final step, examine your factored polynomial to see if the terms in any factors have a common factor. If you find a common factor, factor it out at this point.

Example 6
Factor the following polynomials completely.

   a. $3x^2 - 9x - 12$ **b.** $4a^2y - 36y$

   c. $xz + 2x + yz + 2y$ **d.** $3x^2 - 2x - 16$

Solution

a. $3x^2 - 9x - 12$
First, remove GCF of 3.
$3(x^2 - 3x - 4)$
Then, factor trinomial
$3(x - 4)(x + 1)$

b. $4a^2 y - 36y$
Remove GCF of $4y$.
$4y(a^2 - 9)$
Factor $a^2 - 9 = a^2 - 3^2$ as a difference of 2 squares using $a^2 - b^2 = (a - b)(a + b)$ with $a = a$ and $b = 3$
to get $a^2 - 3^2 = (a - 3)(a + 3)$.
Thus, $4a^2 y - 36y = 4y(a - 3)(a + 3)$.

c. $xz + 2x + yz + 2y$
Since there are four terms, try grouping:
$(xz + 2x) + (yz + 2y) = x(z + 2) + y(z + 2) = (x + y)(z + 2)$

d. $3x^2 - 2x - 16$ (a trinomial)
Multiply $ac = 3(-16) = -48$.
We need factors of -48 whose sum is -2. They are -8 and 6.
Rewrite $3x^2 - 8x + 6x - 16$.
Use grouping: $3x^2 - 8x + 6x - 16$
$$x(3x - 8) + 2(3x - 8)$$
$$(x + 2)(3x - 8)$$

Exercise Set 6.5

Factor each polynomial completely.

1. $x^2 - 64$

2. $4x^2 - 9$

3. $y^4 - 16$

4. $9 - 16x^2 y^4$

5. $64x^2 - 9y^2$

6. $4x^2 - 121$

7. $100 - 4y^2$

8. $x^3 + 1$

9. $a^3 + 27b^3$

10. $64 - y^3$

11. $27 - 8x^3$

12. $1 + 27y^3$

13. $125x^3 + 8z^6$

14. $w^2 + 18w + 45$

15. $9x^2 + 18x$

16. $3x^2 - 6x - 72$

17. $4x^3 + 32x^2 y + 60xy^2$

18. $ab - a + b - 1$

19. $x^2 - 7x + 7x - 49$

20. $5x^2 + 20xy + 3xy + 12y^2$

21. $4x^4 - 9y^4$

22. $27 - 8y^3$ **23.** $16 - 25y^2$ **24.** $x^2 - 2xy - 15y^2$

25. $55p^3 - 20p^2$

Answers to Exercise Set 6.5

1. $(x + 8)(x - 8)$ **2.** $(2x + 3)(2x - 3)$ **3.** $(y + 2)(y - 2)(y^2 + 4)$

4. $(3 - 4xy^2)(3 + 4xy^2)$ **5.** $(8x - 3y)(8x + 3y)$ **6.** $(2x - 11)(2x + 11)$

7. $4(5 - y)(5 + y)$ **8.** $(x + 1)(x^2 - x + 1)$ **9.** $(a + 3b)(a^2 - 3ab + ab^2)$

10. $(4 - y)(16 + 4y + y^2)$ **11.** $(3 - 2x)(9 + 6x + 4x^2)$ **12.** $(1 + 3y)(1 - 3y + 9y^2)$

13. $(5x + 2z^2)(25x^2 - 10xz^2 + 4z^4)$ **14.** $(w + 15)(w + 3)$ **15.** $9x(x + 2)$

16. $3(x - 6)(x + 4)$ **17.** $4x(x + 5y)(x + 3y)$ **18.** $(a + 1)(b - 1)$

19. $(x + 7)(x - 7)$ **20.** $(5x + 3y)(x + 4y)$ **21.** $(2x^2 + 3y^2)(2x^2 - 3y^2)$

22. $(3 - 2y)(9 + 6y + 4y^2)$ **23.** $(4 + 5y)(4 - 5y)$ **24.** $(x + 3y)(x - 5y)$

25. $5p^2(11p - 4)$

6.6   Solving Quadratic Equations Using Factoring

Summary

> Quadratic equations have the form $ax^2 + bx + c = 0$, where a, b, and c are real numbers, $a \neq 0$.

$3x^2 + 10x + 8 = 0$ is an example of a quadratic equation. Notice that the left side of the equation is written in descending powers of x and the right side of the equation is zero.

Summary

> 1. Some quadratic equations can be solved by factoring. Other quadratic equations can be solved using two additional techniques discussed in a later chapter.
>
> 2. The **Zero Factor Property** is used to solve a quadratic equation by factoring.
>
> 3. **Zero Factor Property**
>   **If $ab = 0$ then $a = 0$ or $b = 0$** (In other words, if the product of two algebraic factors is zero, then **at least one of them** must be 0.)

Example 1
Solve $(x - 3)(x + 2) = 0$.

Solution
Since the product of two factors is zero, by the zero factor property, at least one of the factors is zero.

1. Set each factor to zero and solve.
   $$(x - 3) = 0 \quad \text{or} \quad x + 2 = 0$$
   $$x = 3 \qquad\qquad x = -2$$
   The solutions to the equation are –2 and 3.

Example 2
Solve $(2x + 1)(3x - 5) = 0$.

Solution
Use the zero-factor property.

$$2x + 1 = 0 \quad \text{or} \quad 3x - 5 = 0$$
$$2x = -1 \qquad\qquad 3x = 5$$
$$x = -\frac{1}{2} \qquad\qquad x = \frac{5}{3}$$

The solutions are $-\dfrac{1}{2}$ and $\dfrac{5}{3}$.

Summary

To solve a quadratic equation by factoring:
1. Write the equation in standard form with the squared term positive. This will result in one side of the equation being 0.
2. Factor the non-zero side of the equation.
3. Set each factor **containing a variable** equal to zero and solve each equation.
4. Check each solution found in step 3 in the original equation.

Example 3

Solve $x^2 - 4x = 0$.

Solution

1. Set one side equal to 0. (This is done.)

2. Factor non-zero side: $x(x - 4) = 0$

3. Set each factor equal to zero:
   $x = 0$ or $x - 4 = 0$
   $x = 0$ $x = 4$
   The solutions are $x = 0$ and $x = 4$.

4. Check: If $x = 0$, then $0^2 - 4(0) = 0$ or $0 = 0$.   True
   If $x = 4$, then $4^2 - 4(4) = 0$
   $$16 - 16 = 0$$
   $$0 = 0 \quad \text{True}$$

Example 4

Solve $x^2 - 11x = 60$.

Solution:

1. Subtract 60 from each side to make the right side equal to $x^2 - 11x - 60 = 0$.

2. Factor left side to get $(x - 15)(x + 4) = 0$. (Factors of –60 which sum to –11 are –15 and 4.)

3. Set each factor equal to zero:
$$x - 15 = 0 \quad \text{or} \quad x + 4 = 0$$
$$x = 15 \qquad\qquad x = -4$$
The solutions are 15 and –4.

4. Check: If $x = 15$ then $15^2 - 11(15) = 60$
$$225 - 165 = 60$$
$$60 = 60 \text{ True}$$
If $x = -4$, then $(-4)^2 - 11(-4) = 60$
$$16 + 44 = 60$$
$$60 = 60 \text{ True}$$

Example 5
Solve $4x^2 = 40x - 96$.

Solution

1. Set right side equal to zero by adding $-40x + 96$ to both sides. This will result in the coefficient of the squared term being positive.
$$4x^2 - 40x + 96 = 0$$

2. Now, factor out the GCF of 4.
$$4(x^2 - 10x + 24) = 0$$
Factor $x^2 - 10x + 24$ as $(x - 6)(x - 4)$.
So, we now have $4(x - 6)(x - 4) = 0$.

3. Set each factor **containing a variable** equal to zero and solve the equations.
$$x - 6 = 0 \quad \text{or} \quad x - 4 = 0$$
$$x = 6 \qquad\qquad x = 4$$
The solutions are 6 and 4.

4. Check: If $x = 6$, $4(6)^2 = 40(6) - 96$
$$4(36) = 240 - 96$$
$$144 = 144 \text{ True}$$
If $x = 4$, $4(4)^2 = 40(4) - 96$
$$4(16) = 160 - 96$$
$$64 = 64 \text{ True}$$

Example 6

Solve $-x^2 - 7x - 12 = 0$.

Solution

1. Make squared term positive by multiplying both sides by -1.

    $-1(-x^2 - 7x - 12) = 0(-1)$

    or $x^2 + 7x + 12 = 0$

2. Factor the left side.

    $(x + 3)(x + 4) = 0$

3. Set each factor to zero and solve.

    $x + 3 = 0$ or $x + 4 = 0$

       $x = -3$ $x = -4$

    The solutions are -3 and -4.

4. A check will show the solutions are correct.

Example 7

Solve $x^2 = 36$.

Solution

1. Make the right side equal to zero by subtracting 36 from both sides.

    $x^2 - 36 = 0$

2. Factor the left side as a difference of squares using $a^2 - b^2 = (a - b)(a + b)$ with $a = x$ and $b = 6$ to get

    $x^2 - 36 = x^2 - 6^2 = (x - 6)(x + 6)$.

    So, $(x - 6)(x + 6) = 0$.

3. Set each factor to zero and solve.

    $x - 6 = 0$ or $x + 6 = 0$

       $x = 6$ $x = -6$

    The solutions are 6 and -6.

4. A check will show the solutions are correct.

Example 8
The product of 2 consecutive even integers is 224. What are the two integers?

Solution
Let n = first even integer. Then, $n + 2$ is the next consecutive even integer.

We have $n(n + 2) = 224$

$$n^2 + 2n = 224$$

This equation is quadratic so we will set the right hand side of the equation to zero: $n^2 + 2n - 224 = 0$

Factor the left side: $(n + 16)(n - 14) = 0$

Set each factor to zero and solve.
$n + 16 = 0 \qquad n - 14 = 0$
$\quad n = -16 \qquad\quad n = 14$

If $n = -16$, then $n + 2 = -14$.
If $n = 14$, then $n + 2 = 16$.
Thus, our solutions are -16, -14 <u>or</u> 14, 16.

Example 9

The distance, d, that an object falls from rest is given by $d = 16t^2$ where d is measured in feet and t is in seconds. Find the time required for a rock to drop from a cliff which is 256 feet high.

Solution
If $d = 16t^2$ and $d = 256$ ft then $256 = 16t^2$.

1. Subtract 256 from both sides to obtain a zero on the left side. (Remember, we want the coefficient of the squared term to be positive.)
   $$0 = 16t^2 - 256$$

2. Factor out the GCF of 16.
   $$0 = 16(t^2 - 16)$$
   Factor $t^2 - 16$ as a difference of 2 squares.
   $$t^2 - 16 = (t - 4)(t + 4)$$
   Thus, $0 = 16(t - 4)(t + 4)$.

3. Set each factor **containing a variable** equal to zero and solve.
   $t - 4 = 0 \quad$ or $\quad t + 4 = 0$
   $\quad t = 4 \qquad\qquad t = -4$ (not possible)
   Thus, it takes 4 seconds for a rock to drop from a height of 256 feet.

Exercise Set 6.6

Solve each equation.

1. $3y^2 = 8y$ **2.** $a(a-2) = 48$ **3.** $4m^2 + 12m = 0$

4. $6x^2 = -x + 1$ **5.** $4x^2 - 7x = 2$ **6.** $p(p-8) = -4$

7. $6x^2 = 7x + 5$ **8.** $3y^2 = 5y$ **9.** $y^2 - 6y + 5 = 0$

10. $4w(3w+4) = 3$ **11.** $t^2 - 81 = 0$ **12.** $x^2 = 49$

13. $3x^2 + 6x = 0$ **14.** $2x^2 + 5x = 12$ **15.** $2x^2 + x - 1 = 0$

16. $y^2 + 5y - 6 = 0$ **17.** $5x^2 - x = 0$ **18.** $6x^2 = 24$

19. $3x^2 + 14x - 5 = 0$ **20.** $2x^2 - 50 = 0$

21. A right triangle has a hypotenuse of 20 ft. One of the legs of the triangle is 4 feet shorter than the other leg. Find the length of the shorter leg.

22. The product of two consecutive integers is 195. What are the integers?

23. The length of a rectangle is 6 meters less than three times the width. If the area of the rectangle is 144 sq meters, find the dimensions of the rectangle.

24. Find two numbers whose sum is 15 and whose product is 36.

Answers to Exercise Set 6.6

1. $y = 0,\ y = \dfrac{8}{3}$ **2.** $a = -6, a = 8$ **3.** $m = 0, m = -3$

4. $x = \dfrac{1}{3},\ x = -\dfrac{1}{2}$ **5.** $x = 2,\ x = -\dfrac{1}{4}$ **6.** $p = 4$

7. $x = -\dfrac{1}{2},\ x = \dfrac{5}{3}$ **8.** $y = 0,\ y = \dfrac{5}{3}$ **9.** $y = 1, y = 5$

10. $w = \dfrac{1}{6},\ w = -\dfrac{3}{2}$ **11.** $t = 9, t = -9$ **12.** $x = 7, x = -7$

13. $x = -2, x = 0$ **14.** $x = -4,\ x = \dfrac{3}{2}$ **15.** $x = -1,\ x = \dfrac{1}{2}$

16. $y = -6, y = 1$ **17.** $x = 0,\ x = \dfrac{1}{5}$ **18.** $x = 2, x = -2$

19. $x = -5$, $x = \dfrac{1}{3}$ **20.** $x = 5$, $x = -5$ **21.** 12 feet

22. 13, 15 or –13, –15   **23.** 8 m by 18 m   **24.** 3 and 12

Chapter 6 Practice Test

1. Find the greatest common factor of 24 and 80.

2. Find the greatest common factor of $108x^3y^6$ and $24x^4y^2$.

3. Factor by grouping: $xy - 3x + xy - 3c$.

4. Factor completely: $ac - cd - 2a + 2d$.

5. Factor completely: $10p^2q^2 - 25pq^3 + 5pq^2$.

6. Factor completely: $3ay - 9y^2$.

7. Factor completely: $x^2 + 2x - 63$.

8. Factor completely: $x^2 - 14x - 32$.

9. Factor completely: $8x^2 - 10x - 7$.

10. Factor completely: $6y^3 - 36y^2 + 54y$.

11. Factor completely: $50x^3 - 2x^5$.

12. Factor completely: $125 - a^3$.

13. Factor completely: $32 - 200b^2$.

14. Factor completely: $a^3 - 8b^3$.

15. Factor completely: $6w^3 - 54w$.

16. Solve: $3x^2 + 14x = 0$

17. Solve: $x^2 - x = 20$

18. Solve: $3x^2 - 20x = 7$

19. Find two consecutive odd integers whose product is 143.

20. The perimeter of a rectangular rug is 52 feet and its area is 168 ft^2. Find the dimensions of the rug.

Answers to Chapter 6 Practice Test

1. 8

2. $12x^3y^2$

3. $(y-3)(x+c)$

4. $(c-2)(a-d)$

5. $(5pq^2)(2p-5q+1)$

6. $3y(a-3y)$

7. $(x-7)(x+9)$

8. $(x-16)(x+2)$

9. $(4x-7)(2x+1)$

10. $6y(y-3)^2$

11. $2x^3(5-x)(5+x)$

12. $(5-a)(25+5a+a^2)$

13. $8(2+5b)(2-5b)$

14. $(a-2b)(a^2+2ab+4b^2)$

15. $6w(w+3)(w-3)$

16. $-\dfrac{14}{3}, \, 0$

17. $-4, 5$

18. $-7, \dfrac{1}{3}$

19. $-13, -11$ or $13, 11$

20. 12 ft $\times$ 14 ft

Chapter 7

7.1 Simplifying Rational Expressions

Summary

A **rational expression** (also called an **algebraic fraction**) is an algebraic expression of the form $\dfrac{p}{q}$ where p and q are polynomials and $q \neq 0$.

For negative algebraic fraction, $\dfrac{-a}{b} = \dfrac{-a}{b} = \dfrac{a}{-b}$, though we generally do not write a fraction with a negative denominator.

Whenever we have a rational expression containing a variable in the denominator, we always assume that the value or values of the variable that make the denominator zero are excluded.

Example 1
Determine the value or values for which the rational expression is defined.

a. $\dfrac{6x}{10x - 15}$ **b.** $\dfrac{x^2 + 1}{x^2 - 2x - 8}$

Solution

a. $\dfrac{6x}{10x - 15}$

Set the denominator equal to 0 and solve.

$10x - 15 = 0$

$10x = 15$

$x = \dfrac{15}{10} = \dfrac{3}{2}$

So, the rational expression is defined for all values of x except $\dfrac{3}{2}$ $\left(x \neq \dfrac{3}{2}\right)$.

b. $\dfrac{x^2+1}{x^2-2x-8}$

Set the denominator equal to 0 and solve.

$x^2-2x-8=0$

$(x-4)(x+2)=0$

$x-4=0$ or $x+2=0$

     $x=4$ $x=-2$

So, the rational expression is defined for all values of except 4 and –2 ($x\neq4$ and $x\neq-2$).

Summary

To Simplify Rational Expressions:

1. Factor the numerator and denominator as completely as possible.

2. Divide both numerator and denominator by any common factor.

Example 2
Simplify the following.

     a. $\dfrac{2x^3+8x^2-6x}{2x^2}$ **b.** $\dfrac{x^2-4x+3}{x-1}$ **c.** $\dfrac{x^2+x-12}{2x^2+7x-4}$

Solution

     a. $\dfrac{2x^3+8x^2-6x}{2x^2}$

Factor the greatest common factor, $2x$, from each term in numerator. Since $2x$ is a common factor to both numerator and denominator, divide it out.

$$\frac{2x^3+8x^2-6x}{2x^2}=\frac{2x(x^2+4x-3)}{2x^2}=\frac{x^2+4x-3}{x}$$

     b. $\dfrac{x^2-4x+3}{x-1}$

Factor the numerator; then divide out the common factor.

$$\frac{x^2-4x+3}{x-1}=\frac{(x-1)(x-3)}{x-1}=x-3$$

189

c. $\dfrac{x^2 + x - 12}{2x^2 + 7x - 4}$

Factor both the numerator and denominator, then divide out the common factors.

$$\frac{x^2 + x - 12}{2x^2 + 7x - 4} = \frac{(x+4)(x-3)}{(x+4)(2x-1)} = \frac{x-3}{2x-1}$$

d. $\dfrac{x^2 - 7x + 10}{5 - x}$

Factor both numerator and denominator.

$$\frac{x^2 - 7x + 10}{5 - x} = \frac{(x-5)(x-2)}{5 - x}$$

Now, rewrite $5 - x$ as $-1(x - 5)$ so that common factors can be divided out.

$$\frac{(x-5)(x-2)}{5 - x} = \frac{(x-5)(x-2)}{-1(x-5)} = \frac{x-2}{-1} = -1(x-2)$$

Note: A common student error is to try to divide out terms of the numerator and denominator rather than common factors.

$$\frac{x^2 - 3x}{x^2 - 5x} \ne \frac{1-3}{1-5}$$

Exercise Set 7.1

Determine the value or values of the variable when the expression is defined.

1. $\dfrac{3}{4x + 8}$

2. $\dfrac{x - 2}{x^2 - 25}$

3. $\dfrac{x^2}{4x^2 - 8x - 5}$

Reduce each expression to its lowest terms.

4. $\dfrac{4y - 10z}{6}$

5. $\dfrac{x^2 + 3x - 10}{x^2 + 2x - 8}$

6. $\dfrac{x^2 + x - 12}{(x-3)^2}$

7. $\dfrac{x^2 - 3x - 10}{25 - x^2}$

8. $\dfrac{2x^3 + 2x^2 - 4x}{x^3 + 2x^2 - 3x}$

9. $\dfrac{6x^2 - 7x + 2}{6x^2 + 5x - 6}$

10. $\dfrac{2n^2 - 9x + 4}{2n^2 - 5x - 12}$

11. $\dfrac{12 - 6x}{x - 2}$

12. $\dfrac{x^2 - 36}{6 - x}$

13. $\dfrac{x^4 + x^2}{x^2 + 1}$

14. $\dfrac{x^3 + 27}{x + 3}$

15. $\dfrac{x^3 - 8}{x^2 + 2x + 4}$

Answers to Exercise Set 7.1

1. $x \neq -2$

2. $x \neq 5$ and $x \neq -5$

3. $x \neq -\dfrac{1}{2}$ and $x \neq \dfrac{5}{2}$

4. $\dfrac{2y - 5z}{3}$

5. $\dfrac{x + 5}{x + 4}$

6. $\dfrac{x + 4}{x - 3}$

7. $-\dfrac{x + 2}{x + 5}$

8. $\dfrac{2(x + 2)}{x + 3}$

9. $\dfrac{2x - 1}{2x + 3}$

10. $\dfrac{2n - 1}{2n + 3}$

11. -6

12. $-(x + 6)$ or $-x - 6$

13. x^2

14. $x^2 - 3x + 9$

15. $x - 2$

7.2 Multiplication and Division of Rational Expression

Summary

To Multiply Rational Expressions

1. Factor all numerators and denominators completely.

2. Divide out common factors.

3. Multiply numerators together and denominators together.

Example 1

Multiply $\dfrac{8x^2}{9y^3} \cdot \dfrac{3y^2}{4x^3}$.

Solution

$$\frac{8x^2}{9y^3} \cdot \frac{3y^2}{4x^3} = \frac{2 \cdot 2 \cdot 2 \cdot x \cdot x}{3 \cdot 3 \cdot y \cdot y \cdot y} \cdot \frac{3 \cdot y \cdot y}{2 \cdot 2 \cdot x \cdot x \cdot x}$$

$$= \frac{2}{3y} \cdot \frac{1}{x} \quad \text{Divide out common factors}$$

Now, multiply the remaining numerators together and the remaining denominators together.

$$\frac{2}{3y} \cdot \frac{1}{x} = \frac{2}{3xy}$$

Example 2

Multiply $\dfrac{12a^5b}{25x^2y^5} \cdot \dfrac{20xy^2}{16a^3b^3}$.

Solution

$$\frac{12a^5b}{25x^2y^5} \cdot \frac{20xy^2}{16a^3b^3} = \frac{3 \cdot a^2}{5 \cdot x \cdot y^3} = \frac{3a^2}{5xy^3b^2}$$

Example 3

Multiply $\dfrac{8x-12}{14x+7} \cdot \dfrac{42x+21}{32x-48}$.

Solution

$$\frac{8x-12}{14x+7} \cdot \frac{42x+21}{32x-48} = \frac{4(2x-3)}{7(2x+1)} \cdot \frac{21(2x+1)}{16(2x-3)}$$
$$= \frac{3}{4}$$

Example 4

Multiply $\dfrac{x^2+3x}{x^2-3x-4} \cdot \dfrac{x^2-5x+4}{x^2+2x-3}$.

Solution

$$\frac{x^2+3x}{x^2-3x-4} \cdot \frac{x^2-5x+4}{x^2+2x-3} = \frac{x(x+3)}{(x-4)(x+1)} \cdot \frac{(x-4)(x-1)}{(x+3)(x-1)}$$
$$= \frac{x}{x+1}$$

Example 5

Multiply $\dfrac{x^2+x-6}{x^2+7x+12} \cdot \dfrac{x^2+3x-4}{4-x^2}$.

Solution

$$\frac{x^2+x-6}{x^2+7x+12} \cdot \frac{x^2+3x-4}{4-x^2} = \frac{(x+3)(x-2)}{(x+3)(x+4)} \cdot \frac{(x+4)(x-1)}{(2+x)(2-x)}$$
$$= \frac{(x-2)(x-1)}{(2-x)(2+x)}$$
$$= \frac{(x-2)(x-1)}{-1(x-2)(2+x)} \quad \text{Recall: } 2 \cdot x \text{ can be written as } -1(x-2)$$
$$= \frac{(x-1)}{-1(2+x)}$$
$$= -\frac{x-1}{x+2} \text{ or } \frac{-x+1}{x+2}$$

Summary

Division

$$\frac{a}{b} \div \frac{c}{d} = \frac{a}{b} \cdot \frac{d}{c} = \frac{ad}{bc}$$

To Divide Rational Expressions

Invert the divisor (the second fraction) and multiply.

Example 6

Divide $\dfrac{2x^2 y^3}{5a^3 b} \div \dfrac{8xy^5}{25ab^4}$.

Solution

$$\frac{2x^2 y^3}{5a^3 b} \div \frac{8xy^5}{25ab^4} = \frac{2x^2 y^3}{5a^3 b} \cdot \frac{25ab^4}{8xy^5}$$

Example 7

Divide $\dfrac{5a^2 y + 3a^2}{2x^3 + 5x^2} \div \dfrac{10ay + 6a}{6x^3 + 15x^2}$.

Solution

$$\frac{5a^2 y + 3a^2}{2x^3 + 5x^2} \div \frac{10ay + 6a}{6x^3 + 15x^2} = \frac{5a^2 y + 3a^2}{2x^3 + 5x^2} \cdot \frac{6x^3 + 15x^2}{10ay + 6a}$$

$$= \frac{3a}{2}$$

Example 8

Divide $\dfrac{x^2 - x - 2}{x^2 - 7x + 10} \div \dfrac{x^2 - 3x - 4}{40 - 3x - x^2}$.

$$\frac{x^2 - x - 2}{x^2 - 7x + 10} \div \frac{x^2 - 3x - 4}{40 - 3x - x^2} = \frac{x^2 - x - 2}{x^2 - 7x + 10} \cdot \frac{40 - 3x - x^2}{x^2 - 3x - 4}$$

$$\frac{(8 + x)(-1)}{(x - 4)} = \frac{-1(8 + x)}{x - 4} \text{ or } -\frac{x + 8}{x - 4}$$

Exercise Set 7.2

Multiply.

1. $\dfrac{m^3}{2} \cdot \dfrac{4m}{m^4}$

2. $\dfrac{6y^5 x^6}{y^3 x^4} \cdot \dfrac{y^4 x^2}{3y^5 x^7}$

3. $\dfrac{6p^2 q^3}{4p^3 q} \cdot \dfrac{18p^2 q^3}{12p^3 q^3}$

4. $\dfrac{2r}{8r+4} \cdot \dfrac{14r+7}{3}$

5. $\dfrac{6a-10}{5a} \cdot \dfrac{3}{9a-15}$

6. $\dfrac{m^2+6m+9}{m} \cdot \dfrac{m^2}{m^2-9}$

7. $\dfrac{x^2-1}{2x} \cdot \dfrac{1}{1-x}$

8. $\dfrac{6r-5s}{3r+2s} \cdot \dfrac{6r+4s}{5s-6r}$

9. $\dfrac{2r^2+5r-3}{r^2-1} \cdot \dfrac{r^2-2r+1}{2r^2-7r+3}$

10. $\dfrac{a^2-16}{a^2+a-12} \cdot \dfrac{a^2+5a+6}{a^2-2a-8}$

Divide.

11. $\dfrac{6a^4}{a^3} \div \dfrac{12a^2}{a^5}$

12. $\dfrac{25a^2b}{60a^3b^2} \div \dfrac{5a^4b^2}{16a^2b}$

13. $\dfrac{8s^4t^2}{5t^6} \div \dfrac{s^3t^2}{10s^2t^4}$

14. $\dfrac{7k+7}{2} \div \dfrac{4k+4}{5}$

15. $\dfrac{p^2-36}{p+1} \div \dfrac{6-p}{p}$

16. $\dfrac{x^2+3x-40}{x^2+2x-35} \div \dfrac{x^2+2x-48}{x^2+3x-18}$

17. $\dfrac{y^2-y-56}{y^2+8y+7} \div \dfrac{y^2-13y+40}{y^2-4y-5}$

18. $\dfrac{8+2x-x^2}{x^2+7x+10} \div \dfrac{x^2-11x+28}{x^2-x-42}$ \

19. $\dfrac{2x^2-3x-20}{2x^2-7x-30} \div \dfrac{2x^2-5x-12}{4x^2+12x+9}$

20. $\dfrac{9x^2-16}{6x^2-11x+4} \div \dfrac{6x^2+11x+4}{8x^2+10x+3}$

Answers to Exercise Set 7.2

1. 2

2. $\dfrac{2y}{x^3}$

3. $\dfrac{9q^2}{4p^2}$

4. $\dfrac{7r}{6}$

5. $\dfrac{2}{5a}$

6. $\dfrac{m(m+3)}{m-3}$

7. $-\dfrac{x+1}{2x}$

8. -2

9. $\dfrac{(r+3)(r-1)}{(r+1)(r-3)}$

10. $\dfrac{a+3}{a-3}$

11. $\dfrac{a^4}{2}$

12. $\dfrac{4}{3a^3b^2}$

194

13. $\dfrac{16s^3}{t^2}$ **14.** $\dfrac{35}{8}$ **15.** $-\dfrac{p(p+6)}{p+1}$

16. $\dfrac{(x+6)(x-3)}{(x+7)(x-6)}$ **17.** 1   **18.** $-\dfrac{x+6}{x+5}$

19. $\dfrac{2x+3}{x-6}$ **20.** $\dfrac{4x+3}{2x-1}$

7.3 Addition and Subtraction of Rational Expressions with a Common Denominator and Finding the Least Common Denominator

Summary

Addition and Subtraction

$\dfrac{a}{c}+\dfrac{b}{c}=\dfrac{a+b}{c}, \quad c \neq 0$

$\dfrac{a}{c}-\dfrac{b}{c}=\dfrac{a-b}{c}, \quad c \neq 0$

Summary

To Add or Subtract Rational Expressions with a Common Denominator

1. Add or subtract the numerators.

2. Place the sum or differences of the numerator found in step 1 over the common denominator.

3. Simplify the fraction if possible.

Example 1

Add $\dfrac{4}{3t}+\dfrac{9}{3t}$.

Solution

$\dfrac{4}{3t}+\dfrac{9}{3t}=\dfrac{13}{3t}$ (Added numerators)

Cannot be reduced

Example 2

Add $\dfrac{x^2}{x+2}+\dfrac{2x}{x+2}$.

Solution

$$\dfrac{x^2}{x+2}+\dfrac{2x}{x+2}=\dfrac{x^2+2x}{x+2}\quad\text{Add numerators}$$

$$=x\qquad\text{Simplify fraction}$$

Example 3

Add $\dfrac{a^2+a-5}{(a-1)(a+1)}+\dfrac{4-a}{(a-1)(a+1)}$.

Solution

$$\dfrac{a^2+a-5}{(a-1)(a+1)}+\dfrac{4-a}{(a-1)(a+1)}=\dfrac{a^2+a-5+(4-a)}{(a-1)(a+1)}\quad\text{Add numerators}$$

$$=\dfrac{a^2-1}{(a-1)(a+1)}$$

$$=1\qquad\text{Simplify fraction}$$

Example 4

Subtract $\dfrac{3y}{y-2}-\dfrac{6}{y-2}$.

Solution

$$\dfrac{3y}{y-2}-\dfrac{6}{y-2}=\dfrac{3y-6}{y-2}\quad\text{Subtract numerators}$$

$$=\dfrac{3(y-2)}{y-2}$$

$$=3\qquad\text{Simplify fraction}$$

Example 5

Subtract $\dfrac{a^2}{a^2-2a-3} - \dfrac{6a-9}{a^2-2a-3}$.

Solution

$$\frac{a^2}{a^2-2a-3} - \frac{6a-9}{a^2-2a-3} = \frac{a^2-(6a-9)}{a^2-2a-3}$$

$$= \frac{a^2-6a+9}{a^2-2a-3} \quad \text{Subtract numerators}$$

$$= \frac{(a-3)(a-3)}{(a+1)(a-3)}$$

$$= \frac{a-3}{a+1} \quad \text{Simplify fraction}$$

Note: A common student error is to fail to change the signs of all the terms of the numerator being subtracted.

Summary

To Find the Least Common Denominator of Rational Expressions:

1. Factor each denominator completely. Any factor that occurs more than once should be expressed as powers. For example, $(x+5)(x+5)$ should be expressed as $(x+5)^2$.

2. List all different factors (other than 1) that appear in any of the denominators. When the factor appears in more than one denominator, write that factor with the highest power that appears.

3. The least common denominator is the product of all the factors determined in the second step.

Example 6

Find the least common denominator of the following.

a. $\dfrac{2}{3x} + \dfrac{5}{2}$

b. $\dfrac{4a+b}{4a^2b^3} + \dfrac{2a-b^2}{ab^4}$

c. $\dfrac{t}{4t+8} - \dfrac{t^2}{3t+6}$

d. $\dfrac{2x}{x^2+6x+9} + \dfrac{4}{x^2-2x-15}$

Solution

a. $\dfrac{2}{3x} + \dfrac{5}{2}$

The common factors are 3, x and 2. List each factor with its highest power. The least common denominator (LCD) is the product of these factors.

LCD $= 3 \cdot 2 \cdot x = 6x$

b. $\dfrac{4a+b}{4a^2b^3} + \dfrac{2a-b^2}{ab^4}$

The common factors (other than 1) are 4, a, and b. List each factor with its highest power. The LCD is the product of these factors.

LCD $= 4 \cdot a^2 \cdot b^4 = 4a^2b^4$

c. $\dfrac{t}{4t+8} - \dfrac{t^2}{3t+6}$

Here, factor both denominators.

$\dfrac{t}{4(t+2)} - \dfrac{t^2}{3(t+2)}$

Common factors (other than 1) are 4, 3, and $(t+2)$.

LCD $= 4 \cdot 3 \cdot (t+2) = 12(t+2)$

d. $\dfrac{2x}{x^2+6x+9} + \dfrac{4}{x^2-2x-15}$

Here, factor both denominators.

$\dfrac{2x}{(x+3)^2} + \dfrac{4}{(x+3)(x-5)}$

Common factors (other than 1) are $(x+3)$ and $(x-5)$.

List each factor with highest power.

LCD $= (x+3)^2 \cdot (x-5) = (x+3)^2(x-5)$.

Exercise Set 7.3

Add or subtract.

1. $\dfrac{4}{y-3} + \dfrac{6y}{y-3}$

2. $\dfrac{5a+1}{2a+3} + \dfrac{5a+14}{2a+3}$

3. $\dfrac{5x}{18} + \dfrac{7x}{18}$

4. $\dfrac{x}{x^2-1} + \dfrac{1}{x^2-1}$

5. $\dfrac{2x}{x-2} - \dfrac{4}{x-2}$

6. $\dfrac{3x-1}{x^2-5x+4} - \dfrac{2x+3}{x^2-5x+4}$

7. $\dfrac{2n}{3n+4} - \dfrac{5n-3}{3n+4}$

8. $\dfrac{3x-1}{x^2+5x-6} - \dfrac{2x-7}{x^2+5x-6}$

9. $\dfrac{4y+7}{2y^2+7y-4} - \dfrac{y-5}{2y^2+7y-4}$

10. $\dfrac{x+1}{2x^2-5x-12} + \dfrac{x+2}{2x^2-5x-12}$

Find the least common denominator.

11. $\dfrac{x}{2} - \dfrac{y}{x}$

12. $\dfrac{6}{5y} + \dfrac{7}{9y}$

13. $\dfrac{2}{y^2} - \dfrac{5}{y^3}$

14. $\dfrac{5}{t^2} + \dfrac{2}{2t}$

15. $\dfrac{10}{6b^3} + \dfrac{2}{8b^2}$

16. $\dfrac{4}{15a^2b^2} - \dfrac{12}{5ab^4}$

17. $\dfrac{x+1}{4x^2} + \dfrac{x-3}{6x^2 - 12x}$

18. $\dfrac{2x-1}{2x - x^2} - \dfrac{x}{x^2 + x - 6}$

19. $\dfrac{3}{x^2 + x - 2} + \dfrac{x}{x+2}$

20. $\dfrac{x}{x^2 - 16} - \dfrac{4}{x^2 + 8x + 16}$

Answers to Exercise Set 7.3

1. $\dfrac{6y+4}{y-3}$

2. 5

3. $\dfrac{2x}{3}$

4. $\dfrac{1}{x-1}$

5. 2

6. $\dfrac{1}{x-1}$

7. $\dfrac{-3n+3}{3n+4}$

8. $\dfrac{1}{x-1}$

9. $\dfrac{3}{2y-1}$

10. $\dfrac{1}{x-4}$

11. $2x$

12. $45y$

13. y^3

14. $2t^2$

15. $2t^2$

16. $15a^2b^4$

17. $12x^2(x-2)$

18. $x(x-2)(x+3)$

19. $(x+2)(x-1)$

20. $(x+4)^2(x-4)$

7.4   Addition and Subtraction of Rational Expressions

Summary

> **To Add or Subtract Two Rational Expressions with Unlike Denominators:**
>
> 1. Determine the LCD.
>
> 2. Rewrite each fraction as an equivalent fraction with the LCD. This is done by multiplying both the numerator and denominator of each fraction by any factors
>
> 3. Add or subtract the numerators while maintaining the LCD.
>
> 4. When possible, factor the remaining numerator and simplify the fraction.

Example 1

Add $\dfrac{5}{x} + \dfrac{3}{x^2}$.

Solution

(Step 1) $LCD = x^2$

(Step 2) $\dfrac{5}{x} \cdot \dfrac{x}{x} + \dfrac{3}{x^2} = \dfrac{5x}{x^2} + \dfrac{3}{x^2}$

(Step 3) $\dfrac{5x}{x^2} + \dfrac{3}{x^2} = \dfrac{5x+3}{x^2}$

Thus, $\dfrac{5}{x} + \dfrac{3}{x^2} = \dfrac{5x+3}{x^2}$.

Example 2

Subtract $\dfrac{3}{4xy} - \dfrac{1}{2x}$.

Solution

(Step 1) $LCD = 4xy$

(Step 2) $\dfrac{3}{4xy} - \dfrac{1}{2x} \cdot \dfrac{2y}{2y} = \dfrac{3}{4xy} - \dfrac{2y}{4xy}$

(Step 3) $\dfrac{3}{4xy} - \dfrac{2y}{4xy} = \dfrac{3-2y}{4xy}$

Thus, $\dfrac{3}{4xy} - \dfrac{1}{2x} = \dfrac{3-2y}{4xy}$.

Example 3

Add $\dfrac{2x}{x-3} + \dfrac{5}{x+5}$.

Solution

(Step 1)   LCD $= (x-3)(x+5)$

(Step 2) $\dfrac{2x}{x-3} \cdot \dfrac{x+5}{x+5} + \dfrac{5}{x+5} \cdot \dfrac{x-3}{x-3}$

(Step 3) $\dfrac{2x^2 + 10x + 5x - 15}{(x-3)(x+5)} = \dfrac{2x^2 + 15x - 15}{(x-3)(x+5)}$

Thus, $\dfrac{2x}{x-3} + \dfrac{5}{x+5} = \dfrac{2x^2 - 15x - 15}{(x-3)(x+5)}$.

Example 4

Add $\dfrac{x^2}{x-3} + \dfrac{9}{3-x}$.

Solution

(Step 1) $x - 3$ since $3 - x = -1(x-3)$

(Step 2) $\dfrac{x^2}{x-3} + \dfrac{-1}{-1} \cdot \dfrac{9}{3-x} = \dfrac{x^2}{x-3} + \dfrac{-9}{x-3}$

(Step 3) $\dfrac{x^2}{x-3} + \dfrac{-9}{x-3} = \dfrac{x^2 - 9}{x-3}$

(Step 4) $\dfrac{x^2 - 9}{x-3} = \dfrac{(x-3)(x+3)}{(x-3)} = x+3$

Thus, $\dfrac{x^2}{x-3} + \dfrac{9}{3-x} = x+3$.

Example 5

Subtract $\dfrac{x}{x^2-2x-3}-\dfrac{5}{x^2-1}$.

Solution

(Step 1) $\text{LCD} = (x-3)(x-1)(x+1)$

Here, $x^2-2x-3=(x-3)(x+1)$

$\qquad\quad x^2-1=(x-1)(x+1)$

(Step 2) $\dfrac{x}{(x-3)(x+1)}-\dfrac{5}{(x-1)(x+1)}$

$\qquad = \dfrac{x}{(x-3)(x+1)}\cdot\dfrac{x-1}{x-1}-\dfrac{5}{(x-1)(x+1)}\cdot\dfrac{x-3}{x-3}$

(Step 3) $\dfrac{x^2-x}{(x-3)(x+1)(x-1)}+\dfrac{-5x+15}{(x-1)(x+1)(x-3)}$

$\qquad = \dfrac{x^2-6x+15}{(x-3)(x+1)(x-1)}$

Thus, $\dfrac{x}{x^2-2x-3}-\dfrac{5}{x^2-1}=\dfrac{x^2-6x+15}{(x-3)(x+1)(x-1)}$.

Exercise Set 7.4

Add or subtract.

1. $\dfrac{x}{8}-\dfrac{y}{12}$

2. $\dfrac{x+3}{10}+\dfrac{3x-1}{15}$

3. $\dfrac{7}{a}+\dfrac{5}{b}$

4. $\dfrac{4x-3}{6x}+\dfrac{2x+3}{4x}$

5. $\dfrac{2x-3}{2x}+\dfrac{x+3}{3x}$

6. $\dfrac{3y-2}{12y}-\dfrac{y-3}{18y}$

7. $4+\dfrac{5a}{a+3}$

8. $\dfrac{x+3}{6x}-\dfrac{x-3}{8x^2}$

9. $\dfrac{x+5}{3x^2}+\dfrac{2x+1}{2x}$

10. $\dfrac{2}{x-3}+\dfrac{5}{x-4}$

11. $\dfrac{3}{y+6}-\dfrac{4}{y-3}$

12. $\dfrac{3x}{x-4}+\dfrac{2}{x+6}$

13. $\dfrac{2b}{b-7}+\dfrac{5}{7-b}$

14. $\dfrac{4y}{6-y}+\dfrac{5}{y-6}$

15. $\dfrac{x}{x^2-9}+\dfrac{3}{x-3}$

16. $\dfrac{2x}{x^2 - x - 6} - \dfrac{3}{x+2}$

17. $\dfrac{x^2 - 2}{2x^2 - x - 3} - \dfrac{x-2}{2x-3}$

18. $\dfrac{x+2}{2x^2 + 5x + 2} - \dfrac{3}{2x+1}$

Answers to Exercise Set 7.4

1. $\dfrac{3x - 2y}{24}$

2. $\dfrac{9x + 7}{30}$

3. $\dfrac{7b + 5a}{ab}$

4. $\dfrac{14x + 3}{12x}$

5. $\dfrac{8x - 3}{6x}$

6. $\dfrac{7}{36}$

7. $\dfrac{3(3a + 4)}{a + 3}$

8. $\dfrac{4x^2 + 9x + 9}{24x^2}$

9. $\dfrac{6x^2 + 5x + 10}{6x^2}$

10. $\dfrac{7x - 23}{(x-3)(x-4)}$

11. $\dfrac{-y - 33}{(y+6)(y-3)}$

12. $\dfrac{3x^2 + 20x - 8}{(x-4)(x+6)}$

13. $\dfrac{2b - 5}{b - 7}$

14. $\dfrac{-4y + 5}{y - 6}$

15. $\dfrac{4x + 9}{(x+3)(x-3)}$

16. $\dfrac{-x + 9}{(x-3)(x+2)}$

17. $\dfrac{x}{(2x-3)(x+1)}$

18. $-\dfrac{2}{2x+1}$

7.5   Complex Fractions

Summary

A complex fraction is one that has a fraction in its numerator or its denominator or in both its numerator and denominator.

$\dfrac{\dfrac{a+b}{a}}{\dfrac{a-b}{b}}$ numerator of complex fraction

------ <----------------------->

          main fraction line

          denominator of complex fraction

Summary

> **To Simplify a Complex Fraction by Combining Terms**
>
> 1. Add or subtract the fractions in both the numerator and denominator of the complex fraction to obtain single fractions in both the numerator and the denominator.
>
> 2. Invert the denominator of the complex fraction and multiply the numerator by it.
>
> 3. Simplify further if possible.

Example 1

Simplify $\dfrac{\frac{2}{3} - \frac{1}{8}}{\frac{3}{4} + \frac{1}{3}}$.

Solution

(Step 1) $\quad \dfrac{\frac{2}{3} \cdot \frac{8}{8} - \frac{1}{8} \cdot \frac{3}{3}}{\frac{3}{4} \cdot \frac{3}{3} + \frac{1}{33} \cdot \frac{4}{4}}$ $\begin{aligned} \text{LCD} &= 8 \cdot 3 = 24 \\ \text{LCD} &= 4 \cdot 3 = 12 \end{aligned}$

$\qquad\quad = \dfrac{\frac{16}{24} - \frac{3}{24}}{\frac{9}{12} + \frac{4}{12}} = \dfrac{\frac{13}{24}}{\frac{13}{12}}$

(Step 2) $\quad = \dfrac{13}{24} \cdot \dfrac{12}{13} = \dfrac{13}{24} \cdot \dfrac{12}{13} = \dfrac{1}{2}$

Example 2

Simplify $\dfrac{1 - \frac{4}{x^2}}{1 - \frac{2}{x}}$.

Solution

(Step 1) $\quad \dfrac{\frac{1}{1} \cdot \frac{x^2}{x^2} - \frac{4}{x^2}}{\frac{1}{1} \cdot \frac{x}{x} - \frac{2}{x}}$ $\begin{aligned} \text{LCD} &= x^2 \\ \text{LCD} &= x \end{aligned}$

$\qquad\quad = \dfrac{\frac{x^2}{x^2} - \frac{4}{x^2}}{\frac{x}{x} - \frac{2}{x}} = \dfrac{\frac{x^2 - 4}{x^2}}{\frac{x - 2}{x}}$

(Step 2) $\quad = \dfrac{x^2 - 4}{x^2} \cdot \dfrac{x}{x - 2} = \dfrac{x(x-2)(x+2)}{x^2(x-2)} = \dfrac{x+2}{x}$

Summary

+---+
| **To Simplify a Complex Fraction Using** |
| **Multiplication First:** |
| |
| 1. Find the least common denominator of **all** the |
| denominators appearing in the complex |
| fractions. |
| |
| 2. Multiply both the numerator |
| |
| 3. Simplify when possible. |
+---+

Example 3

Simplify $\dfrac{\frac{2}{3}-\frac{1}{8}}{\frac{3}{4}+\frac{1}{3}}$.

Solution

(Step 1)   LCD of all denominators $= 2^3 \cdot 3 = 24$

(Step 2) $\dfrac{24}{24} \cdot \dfrac{\left(\frac{2}{3}-\frac{1}{8}\right)}{\left(\frac{3}{4}+\frac{1}{3}\right)} = \dfrac{(24)\left(\frac{2}{3}\right)-(24)\left(\frac{1}{8}\right)}{(24)\left(\frac{3}{4}\right)+(24)\left(\frac{1}{3}\right)} = \dfrac{16-3}{18+8} = \dfrac{13}{26}$

(Step 3) $\dfrac{13}{26} = \dfrac{1}{2}$

Example 4

Simplify $\dfrac{1-\frac{4}{x^2}}{11-\frac{2}{x}}$.

Solution

(Step 1)   LCD of all denominators $= x^2$

(Step 2) $\dfrac{x^2}{x^2} \dfrac{\left(1-\frac{4}{x^2}\right)}{\left(1-\frac{2}{x}\right)} = \dfrac{x^2(1)-x^2\left(\frac{4}{x^2}\right)}{x^2(1)-x^2\left(\frac{2}{x}\right)} = \dfrac{x^2-4}{x^2-2x}$

(Step 3) $= \dfrac{(x+2)(x-2)}{x(x-2)} = \dfrac{x+2}{x}$

Exercise Set 7.5

Simplify each expression.

1. $\dfrac{2 - \frac{2}{3}}{3 + \frac{1}{6}}$

2. $\dfrac{\frac{1}{5} - \frac{1}{2}}{\frac{2}{5} + \frac{3}{10}}$

3. $\dfrac{5 - \frac{2}{4}}{3 + \frac{1}{3}}$

4. $\dfrac{\frac{3a^2 b}{2c^3}}{\frac{4ab^4}{6c^4}}$

5. $\dfrac{\frac{2t^2 u^4}{5r^4}}{\frac{8tu^2}{15r}}$

6. $\dfrac{1 + \frac{3}{x}}{1 - \frac{4}{y}}$

7. $\dfrac{\frac{1}{x} + 2}{\frac{1}{x} + 3}$

8. $\dfrac{\frac{1}{a} - \frac{2}{B}}{\frac{3}{a} - \frac{5}{b}}$

9. $\dfrac{\frac{x+1}{x-1}}{\frac{x+1}{x}}$

10. $\dfrac{\frac{b-1}{b}}{\frac{b^2 - 1}{4}}$

11. $\dfrac{\frac{x}{y} - \frac{y}{x}}{\frac{x}{y} + \frac{y}{x}}$

12. $\dfrac{\frac{x}{y} - \frac{y}{x}}{\frac{y}{x} - \frac{x}{y}}$

Answers to Exercise Set 7.5

1. $\dfrac{14}{19}$

2. $-\dfrac{3}{7}$

3. $\dfrac{27}{20}$

4. $\dfrac{9ac}{4b^3}$

5. $\dfrac{3tu^2}{4r^3}$

6. $\dfrac{xy + 3y}{xy - 4x}$

7. $\dfrac{1 + 2x}{1 + 3x}$

8. $\dfrac{b - 2a}{3b - 5a}$

9. $\dfrac{x}{x - 1}$

10. $\dfrac{4}{b(b+1)}$

11. $\dfrac{x^2 - y^2}{x^2 + y^2}$

12. -1

7.6 Solving Rational Equations

Summary

To Solve Rational Equations:

1. Determine the LCD of all fractions in the equation.

2. Multiply **both** sides of the equation by the LCD. **This will result in every term in the equation being multiplied by the LCD.**

3. Remove any parentheses and combine like terms on each side of the equation.

4. Solve the equations using the properties discussed in earlier chapters.

5. Check your solution in the original equation.

Example 1

Solve $\dfrac{x}{2} - \dfrac{x}{3} = 10$.

Solution

(Step 1) $\text{LCD} = 2 \cdot 3 = 6$

(Step 2) $6\left(\dfrac{x}{2} - \dfrac{x}{3} \right) = 6(10)$

(Step 3) $\left(6 \cdot \dfrac{x}{2} \right) - \left(6 - \dfrac{x}{3} \right) = 6(10)$

(Step 4) $3x - 2x = 60$

                $x = 60$

(Step 5) $\dfrac{x}{2} - \dfrac{x}{3} = 10$ Substitute $x = 60$

             $\dfrac{60}{2} - \dfrac{60}{3} = 10$

                $30 - 20 = 10$

                      $10 = 10$ True

          The solution is 60.

Example 2

Solve $\dfrac{3x+1}{4} = \dfrac{13-x}{2}$.

Solution

(Step 1)   LCD $= 4$

(Step 2) $4\left(\dfrac{3x+1}{4}\right) = 4\left(\dfrac{13-x}{2}\right)$

(Step 3) $4 \cdot \dfrac{3x+1}{4} = 4 \cdot \dfrac{13-x}{2}$

$3x + 1 = 2(13 - x)$

$3x + 1 = 26 - 2x$

(Step 4) $3x + 1 = 26 - 2x$

$5x + 1 = 26$

$5x = 25$

$x = 5$

(Step 5) $\dfrac{3x+1}{4} = \dfrac{13-x}{2}$ Substitute $x = 5$

$\dfrac{3(5)+1}{4} = \dfrac{13-5}{2}$

$\dfrac{15+1}{4} = \dfrac{8}{2}$

$\dfrac{16}{4} = \dfrac{8}{2}$

$4 = 4$ True

The solution is 5.

Example 3

Solve $\dfrac{x+9}{2} = \dfrac{15}{x} - 3$.

Solution

(Step 1)   LCD $= 2x$

(Step 2) $2x\left(\dfrac{x+9}{2x}\right) = 2x\left(\dfrac{15}{x} - 3\right)$

(Step 3) $2x \cdot \dfrac{x+9}{2x} = 2x \cdot \dfrac{15}{x} - 2x(3)$

$x + 9 = 30 - 6x$

(Step 4) $7x + 9 = 30$

$$7x = 21$$

$$x = 3$$

(Step 5) $\dfrac{x+9}{2x} = \dfrac{15}{x} - 3$ Substitute $x = 3$,

$$\dfrac{3+9}{2(3)} = \dfrac{15}{3} - 3$$

$$\dfrac{12}{6} = 5 - 3$$

$$2 = 2 \text{ True}$$

The solution is 3.

Example 4

Solve $\dfrac{x}{x+2} + \dfrac{2}{x-2} = \dfrac{x+6}{x^2-4}$.

Solution

(Step 1) $\text{LCD} = (x+2)(x-2)$

(Step 2) $(x+2)(x-2)\left(\dfrac{x}{x+2} + \dfrac{2}{x-2}\right) = (x+2)(x-2)\left(\dfrac{x+6}{x^2-4}\right)$

(Step 3) $(x+2)(x-2)\left(\dfrac{x}{x+2}\right) + (x+2)(x-2)\left(\dfrac{2}{x-2}\right) = (x+2)(x-2)\left(\dfrac{x+6}{x^2-4}\right)$

$$x(x-2) + 2(x+2) = x+6$$

$$x^2 - 2x + 2x + 4 = x + 6$$

$$x^2 + 4 = x + 6$$

(Step 4) $x^2 - x - 2 = 0$

$$(x-2)(x+1) = 0$$

$$x - 2 = 0 \quad \text{or} \quad x + 1 = 0$$

$$x = 2 \qquad\quad x = -1$$

(Step 5) Check $x = -1$.

$$\frac{x}{x+2} + \frac{2}{x-2} = \frac{x+6}{x^2-4}$$

$$\frac{-1}{-1+2} + \frac{2}{-1-2} = \frac{-1+6}{(-1)^2-4}$$

$$-\frac{1}{1} + \frac{2}{-3} = \frac{5}{1-4}$$

$$-1 + \left(-\frac{2}{3}\right) = \frac{5}{1-4}$$

$$-1 + \left(-\frac{2}{3}\right) = -\frac{5}{3}$$

$$-\frac{5}{3} = -\frac{5}{3}$$

Check $x = 2$.

$$\frac{x}{x+2} + \frac{2}{x-2} = \frac{x+6}{x^2-4}$$

$$\frac{2}{2+2} + \frac{2}{2-2} = \frac{2+6}{2^2-4}$$

$$\frac{2}{4} + \frac{2}{0} = \frac{8}{0} \qquad \left(\frac{2}{0} \text{ and } \frac{8}{0} \text{ are not real numbers.}\right)$$

Therefore, 2 is extraneous.
The only solution is -1.

Exercise Set 7.6

Solve each equation and check your solution.

1. $\dfrac{6}{2a+1} = 2$

2. $\dfrac{9}{2x-5} = -2$

3. $\dfrac{2x}{3} - \dfrac{5}{2} = -\dfrac{1}{2}$

4. $\dfrac{x}{3} - \dfrac{1}{4} = \dfrac{1}{12}$

5. $\dfrac{2x-5}{8} + \dfrac{1}{4} = \dfrac{x}{8} + \dfrac{3}{4}$

6. $\dfrac{3x+4}{12} - \dfrac{1}{3} = \dfrac{5x+2}{12} - \dfrac{1}{2}$

7. $2 + \dfrac{5}{x} = 7$

8. $3 + \dfrac{8}{n} = 5$

9. $\dfrac{3}{x-2} = \dfrac{4}{x}$

10. $\dfrac{5}{x+3} = \dfrac{3}{x-1}$

11. $2 + \dfrac{3}{a-3} = \dfrac{a}{a-3}$

12. $\dfrac{2x}{x+4} = \dfrac{3}{x-1}$

13. $x + \dfrac{6}{x-2} = \dfrac{3x}{x-2}$

14. $\dfrac{x+2}{x^2-x-2} + \dfrac{x+1}{x^2-4} = \dfrac{1}{x+1}$

Answers to Exercise Set 7.6

1. 1

2. $\dfrac{1}{4}$

3. 3

4. 1

5. 9

6. 2

7. 1

8. 4

9. 8

10. 7

11. no solution

12. $-\dfrac{3}{2}$, 4

13. −3

7.7   Rational Equations: Applications and Problem Solving

Many applications of algebra involve rational equations. After we set up the application as an equation, we solve the rational equation as we did in Section 6.6.

Example 1
One number is three times the other. If the sum of the reciprocal of the number is 12, find the number.

Solution
Let x = one number,
then $3x$ = other number

$$\begin{pmatrix}\text{reciprocal} \\ \text{of one numer}\end{pmatrix} + \begin{pmatrix}\text{reciprocal} \\ \text{of the other}\end{pmatrix} = 12$$

$$\text{or } \frac{1}{x} + \frac{1}{3x} = 12$$

$$3x\left(\frac{1}{x} + \frac{1}{3}x\right) = 3x(12) \text{ Multiply by } 3x$$

$$3 + 1 = 36x$$

$$4 = 36x$$

$$\frac{4}{36} = x$$

$$\frac{1}{9} = x$$

One number is $\dfrac{1}{9}$ and the other number is $3\left(\dfrac{1}{9}\right) = \dfrac{3}{9} = \dfrac{1}{3}$.

Example 2
The area of a triangle is 20 square feet. If the base is 3 feet longer than the height, find the base and height.

Solution
Let x = height,
then $x + 3$ = base

$$\text{Area} = \frac{1}{2} \cdot \text{base} \cdot \text{height}$$

$$20 = \frac{1}{2}(x+3)(x)$$

$$2 \cdot 20 = 2 \cdot \frac{1}{2}(x+3)(x)$$

$$40 = (x+3)(x)$$

$$40 = x^2 + 3x$$

$$0 = x^2 + 3x - 40$$

$$0 = (x+8)(x-5)$$

$$x + 8 = 0 \quad \text{or} \quad x - 5 = 0$$

$$x \quad = -8 \qquad x \quad = 5$$

Since dimensions cannot be negative, eliminate –8. Therefore, height is 5 feet and base is 5 + 3 = 8 feet.

Example 3
A boat travels downstream a distance of 20 miles in the same time it takes to travel upstream a distance of 16 miles. If the current has a speed of 4 mph, find the speed of the boat in still water.

Solution
Let x = rate in still water

So $x - 4$ = rate upstream

and $x + 4$ = rate downstream

	Distance	Rate	Time
Upstream	16	$x - 4$	$\dfrac{16}{x - 4}$
Downstream	20	$x + 4$	$\dfrac{20}{x + 4}$

$$\text{time} = \frac{\text{distance}}{\text{rate}}$$

Time upstream = Time downstream

$$\frac{16}{x-4} = \frac{20}{x+4}$$

$$16(x+4) = 20(x-4)$$

$$16x + 64 = 20x - 80$$

$$-4x + 64 = -80$$

$$-4x = -144$$

$$x = 36$$

Thus, the speed of the boat is 36 mph.

Summary

Problem in which two or more machines or people work together to complete a certain task are sometimes referred to as **work problems.** We represent the total amount of work done by the number one which represents one whole job completed.

Part of task done by first person or machine + part of task done by second person or machine = 1(one whole task completed)

To determine the part of the task completed by each person or machine, we use the formula:

Part of task completed = rate · time

Example 4
An older pump can empty a swimming pool in 60 hours. A newer pump requires 40 hours to empty the swimming pool. If both pumps are used together, how long will it take to empty the swimming pool?

Solution
Set up a table as follows:

	Rate of Work	Time worked	Part of Task
Old pump:	$\dfrac{1}{60}$	t	$\dfrac{t}{60}$
New pump:	$\dfrac{1}{40}$	t	$\dfrac{t}{40}$

$$\begin{pmatrix} \text{Part of pool} \\ \text{emptied by} \\ \text{old pump} \end{pmatrix} + \begin{pmatrix} \text{part of pool emptied} \\ \text{by new pump} \end{pmatrix} = 1 \ (\text{entire pool emptied})$$

$$\frac{t}{60} \quad + \quad \frac{t}{40} \qquad = 1$$

$$120\left(\frac{t}{60} + \frac{t}{40}\right) \qquad = (120)(1)$$

$$2t \quad + \quad 3t \qquad = 120$$

$$5t \qquad = 120$$

$$t \qquad = 24$$

It takes 24 hours to empty the pool if both pumps work together.

Exercise Set 7.7

For each problem

   a. write an equation that can be used to solve the problem.

   b. solve the problem.

  1. The base of a triangle is 5 meters longer than its height. Find the base and height of the triangle if the area is 7 square meters.

  2. The height of triangle is 3 less than twice its base. Find the base and height of the triangle if the area is 27 square units.

  3. One number is twice a second number. Find the number if the sum of the reciprocals is 12.

  4. The numerator of fraction $\dfrac{2}{3}$ is increased by an amount so that the value of the resulting number is 5. Find the amount that the numerator was increased.

  5. The speed of a boat in still water is 20 mph. The boat traveled 75 miles down a river in the same amount of time as it traveled 45 miles up the river. Find the rate of the river's current.

  6. An express train travels 300 miles in the same amount of time that a freight train travels 180 miles. The rate of the express train is 20 mph faster than the freight train. Find the rate of each train.

7. If one person can mow a lawn in 20 minutes and a second person requires 30 minutes to mow it, how long would it take both people to mow the lawn when they work together?

8. A pipe can fill a swimming pool three times faster than a second pipe. With both pipes working together, they can fill the pool in 9 hours. How long does it take for each of the pipes working separately to fill the pool?

9. A mason can construct a wall in 10 hours. With a helper, the task would take 6 hours. How long would it take the helper working alone to construct the wall?

Answers to Exercise Set 7.7

1. height = 2 meters

    base = 7 meters

2. base = 6 units

    height = 9 units

3. one number = $\dfrac{1}{4}$

    other number = $\dfrac{1}{8}$

4. amount = 13

5. rate of current = 5 mph

6. freight train = 30 mph
    express train = 50 mph

7. 12 minutes

8. 12 hours, 36 hours

9. 15 hours

7.8 Variation

Summary

Direct Variation

If a variable y varies directly as x, then $y = kx$, where k is the **constant of proportionality**. The two related variables will both **increase** together or both **decrease** together.

Example 1
The property tax, T, varies directly as the assessed value, v, of the home. Write a variation equation for this statement.

Solution
Since T = property tax and v = assessed value, the equation is $T = kv$, where k is the constant of proportionality.

Example 2
Suppose y varies directly as x and $y = 5$ when $x = 4$.

Solution
If y varies directly as x, then $y = kx$, where k is the constant of proportionality. If $y = 5$ when $x = 4$, then $5 = k(4)$. Thus, $k = \dfrac{5}{4}$. We now have $y = \dfrac{5}{4}x$. If $x = 20$, then $y = \dfrac{5}{4}(20) = 25$.

Summary

> ### Inverse Variation
>
> If a variable y varies inversely with a variable x, then
> $y = \dfrac{k}{x}$ (or $xy = k$), where k is the constant of
> proportionality. This means that as one quantity
> **increases** the other quantity **decreases** or vice versa.

Example 3
The time to do a job, T, varies inversely as the number of workers, n. If it takes 7 hours for 3 workers to do a certain job, how long should it take 4 workers to do the same job?

Solution
We have $T = \dfrac{k}{n}$, since T varies inversely as n. Substitute the given values into the equation: $7 = \dfrac{k}{3}$, or $k = 21$.

Our new equation is $T = \dfrac{21}{n}$. If $n = 4$, then $T = \dfrac{21}{4} = 5.25$. It should take 4 workers 5.25 hours to do the same job.

Summary

> ### Joint Variation
>
> The general form a **joint variation** where y varies
> directly as x and z is
>
> $y = kxz$, where k is the constant of proportionality.

The following examples illustrate combined variations.

Example 4
V varies jointly as l, w and h and inversely as n; $V = 5$ when $l = 2$, $w = 3$, $h = 4$, and $n = 6$. Find V when $l = 4$, $w = 3$, $h = 1$, and $n = 5$.

Solution
$V = \dfrac{klwh}{n}$

Substitute the values to get $5 = \dfrac{k(2)(3)(4)}{6}$ or $5 = 4k$. Thus, $\dfrac{5}{4} = k$.

Now use this value of k and the new values of the four variables to find V: $V = \dfrac{\frac{5}{4}(4)(3)(1)}{5}$ or $V = 3$.

Example 5

The electrical resistance of a wire, R, varies directly as its length, L, and inversely as its cross-sectional area, A. If the resistance of a wire is 0.2 ohm when the length is 200 feet and its cross-sectional area is 0.05 square inches, find the resistance of a wire whose length is 5000 feet with a cross-sectional area of 0.01 square inches.

Solution

Use the equation $R = \dfrac{kL}{A}$. Now, substitute the first set of values into this equation: $0.2 = \dfrac{k(200)}{0.05}$.

Now solve for k to get $0.2 = 4000k$ or $k = \dfrac{0.2}{4000} = 5 \times 10^{-5}$. Now, substitute the second set of values into the

equation: $R = \dfrac{(5 \times 10^{-5})(5000)}{0.01} = 25$ ohms.

Exercise Set 7.8

Write the variation equation and find the quantities indicated.

1. x varies directly as y. Find x when $y = 12$ and $k = 6$.

2. T varies directly as the square of D and inversely as F. Find T when $D = 8$, $F = 15$ and $k = 12$ (the constant of proportionality).

3. y varies inversely as x, and $y = 2$ when $x = 7$. Find y when $x = 5$.

4. y varies directly as x and inversely as z^2; $y = 0.5$ when $x = 8$ and $z = 2$. Find y when $x = 4$ and $z = -1$.

5. F varies directly as q_1 and q_2 and inversely as the square of d. If $F = 8$ when $q_1 = 2$, $q_2 = 8$ and $d = 4$, find F when $q_1 = 28$, $q_2 = 12$ and $d = 2$.

6. The wattage rating, W, of an appliance varies jointly as the square of the current I and the resistance R. If the wattage is 1 watt when the current is 0.1 amperes and the resistance is 100 ohms, find the wattage when the current is 0.5 amperes and the resistance is 500 ohms.

7. The volume, V, of a cylinder varies jointly as the square of the radius r of the cylinder and the height h of the cylinder. If the volume of a cylinder is $\dfrac{400\pi}{3}$ cubic inches when the radius is 5 inches and the height is 4 inches, find the volume of the cylinder with radius 10 inches and height 4 inches.

8. W varies jointly as A and h and inversely as d^2; $w = 1$ when $A = 9$, $h = 2$, and $d = 6$. Find w when $A = 9$, $h = 2$ and $d = 2$.

9. The weight (w) of a solid cube varies directly as the cube of an edge (e). If a cube with edge 8 inches weighs 10 pounds, what will a cube with edge 7 inches made of the same material weigh? Round to the nearest tenth of a pound.

10. Write the relationship algebraically: w varies jointly as s and t and inversely as the square of r.

Answers to Exercise Set 7.8

1. 72

2. 51.2 or $\dfrac{256}{5}$

3. 2.8 or $\dfrac{14}{5}$

4. 1

5. 672

6. 25 watts

7. $\dfrac{800\pi}{3}$ cu in.

8. 9

9. 6.7 pounds

10. $w = \dfrac{kst}{r^2}$

Chapter 7 Practice Test

Perform the operations indicated.

1. $\dfrac{2x^2y}{7z^3} \cdot \dfrac{21z^4}{8x^3y^5}$

2. $\dfrac{x^5y^3}{x^2-x-6} \cdot \dfrac{x^2-9}{x^2y^4}$

3. $\dfrac{x^2+3x+2}{x^2+5x+4} \div \dfrac{x^2-x-6}{x^2+2x-15}$

4. $\dfrac{x^2-x-56}{x^2+8x+7} \div \dfrac{x^2-13x+40}{x^2-4x-5}$

5. $\dfrac{2x-5}{3x} + \dfrac{x+5}{3x}$

6. $\dfrac{2x}{x^2+3x-10} - \dfrac{4}{x^2+3x-10}$

7. $\dfrac{3}{a^2} - \dfrac{5}{a^3}$

8. $5 + \dfrac{2a}{a-5}$

9. $\dfrac{3}{x^2-4} + \dfrac{x}{(x-2)^2}$

Simplify each expression.

10. $\dfrac{5 - \frac{3}{4}}{6 + \frac{1}{8}}$

11. $\dfrac{a + \frac{1}{a^2}}{\frac{1}{2a}}$

12. $\dfrac{2x}{3} + \dfrac{4x}{5} = 4$

13. $\dfrac{x}{x+4} = \dfrac{11}{x^2-16} + 2$

Solve the problems.

14. A small plane can fly at 110 mph in calm air. Flying with the wind, the plane can fly 260 miles in the same amount of time it can fly 180 miles against the wind. Find the rate of the wind.

15. One pipe can fill a tank in 9 hours, while a second pipe requires 18 hours to fill the tank. How long would it take both pipes working together to fill the tank?

16. *P* varies directly as *Q* and inversely as *R*. If $P = 8$ when $Q = 4$ and $R = 10$, find *P* when $Q = 10$ and $R = 20$.

Answers to Chapter 7 Practice Test

1. $\dfrac{3z}{4xy^4}$

2. $\dfrac{x^3(x+3)}{y(x+2)}$

3. $\dfrac{x+5}{x+4}$

4. 1

5. 1

6. $\dfrac{2}{x+5}$

7. $\dfrac{3a-5}{a^3}$

8. $\dfrac{7a-25}{a-5}$

9. $\dfrac{x^2+5x-6}{(x-2)^2(x+2)}$

10. $\dfrac{37}{49}$

11. $\dfrac{2a^3+2}{a}$

12. $\dfrac{30}{11}$

13. $-7, 3$

14. 20 mph

15. 6 hours

16. 10

Chapter 8

8.1 More on Graphs

Summary

Cartesian Coordinate System

1. The **Cartesian coordinate system** consists of two axes (number lines) which meet at right angles at a point called the **origin**.

2. To graph a point, its **x-coordinate** and **y-coordinate** must be known.

3. An **ordered pair** (x, y) is used to give the coordinates of a point. The x-coordinate is always listed first.

Example 1
Plot the following ordered pairs: **a.** $(-2, 3)$ **b.** $(4, 0)$ **c.** $(0, -3)$ **d.** $(5, 8)$

Solution

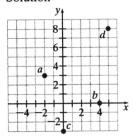

Summary

A **graph** is an illustration of the set of points whose coordinates satisfy the equation. Sometimes when drawing a graph, we list a few points that satisfy the equation in a table and then plot the points. We then draw a line through the points to obtain the graph.

Example 2

Graph $y = \frac{1}{2}x + 1$.

Solution
Construct a table, plot the points, and draw the graph.

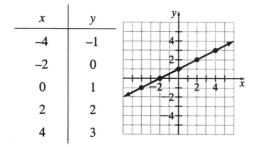

x	y
-4	-1
-2	0
0	1
2	2
4	3

Example 3

Graph $y = x^2 - 1$.

Solution
Construct a table, plot the points, and draw the graph.

x	y
-3	8
-2	3
-1	0
0	-1
1	0
2	3
3	8

Example 4
Graph $y = |x| - 2$.

Solution
Construct a table, plot the points, and draw the graph.

x	y
–4	2
–3	1
–2	0
–1	–1
0	–2
1	–1
2	0
3	1
4	2

Exercise Set 8.1

1. Plot the following points.

    $A(3, 2)$ $B(-2, 4)$ $C(-4, -1)$

    $D(5, -2)$ $E(-3, 0)$

2. Graph the following.

    a. $y = 2x + 1$ **b.** $y = \dfrac{1}{3}x$ **c.** $y = x^2 + 1$

    d. $y = \dfrac{2}{x}$ **e.** $y = |x| + 3$

Answers to Exercise Set 8.1

1.

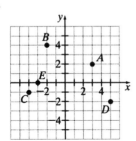

2. a.

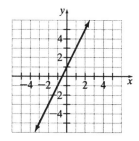

b.

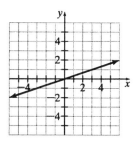

c.

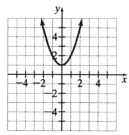

d.

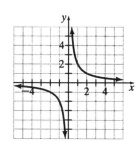

e.

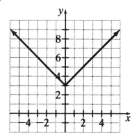

8.2 Functions

Summary

Terminology of Sets

1. A **set** is a collection of objects.

2. Objects in the set are called **elements** of the set.

3. When the elements of a set are listed within braces, the set is said to be in **roster form**.

4. The symbol ∈ represents "is a member of" or "belongs to."

5. If the elements of a set can be counted, the set is a **finite set**.

6. If it is impossible to list all elements of a set, the set is called **infinite.**

Example 1

The set of **natural numbers**, $N = \{1, 2, 3, 4, 5 \ldots\}$, the set of whole numbers, $W = \{0, 1, 2, 3, \ldots\}$, and the set of integers, $I = \{\ldots, -3, -2, -1, 0, 1, 2, 3, \ldots\}$ are infinite sets since it is impossible to list all the elements of each set.

Example 2
The set $\{1, 5, 9, 17\}$ is a finite set having four elements.

Subsets, Null set

1. The **null set** or **empty set** is the set which contains **no** elements. It is symbolized as { } or $\emptyset$.

2. Set A is a **subset** of set B, symbolized by $A \subseteq B$, if every element of A is also an element of B.

Example 3
The set $\{a, c, d\}$ is a subset of $\{a, b, c, d, e, f\}$ since every element of the first set is an element of the second set.

Example 4
The set $\{1, 2, 5, 9\}$ is not a subset of $\{1, 2, 4, 6, 9, 10\}$ since the element 5 from the first set is not an element of the second set.

Example 5
Find the set of **whole number** solutions of $x + 2 = 0$.

Solution
Since $x = -2$ is the only solution of $x + 2 = 0$, and $x = -2$ is **not** a whole number (it is an integer), the solution set of $x + 2 = 0$ in the set of whole numbers is { } (null set).

Summary

Greater than, Less than

1. The symbol ">" is read "is greater than."

2. The symbol "$\geq$" is read "is greater than or equal to."

3. The symbol "<" is read "is less than."

4. The symbol "$\leq$" is read "is less than or equal to."

5. If $a > b$, then a lies to the right of b on the real number line.

6. If a is not equal to b, then either $a > b$ or $a < b$. Here, we write $a \neq b$.

Example 6

Insert > or < symbol between the numbers to make a true statement.

    a. 10   7   **b.** −17   −23   **c.** −5   0

Solution:

    a. $10 > 7$ since 10 lies to the right of 7 on the real number line.

    b. $-17 > -23$ since −17 lies to the right of −23 on the number line.

    c. $-5 < 0$ since −5 lies to the left of 0 on the number line.

Example 7

Illustrate the set $\{x \mid -2 \le x < 3 \text{ and } x \in I\}$ on a number line.

Solution

$$-4 \quad -3 \quad -2 \quad -1 \quad 0 \quad 1 \quad 2 \quad 3 \quad 4$$

Example 8

Illustrate the set $\{x \mid -2 \le x < 3\}$ on a number line.

Solution

$$-4 \quad -3 \quad -2 \quad -1 \quad 0 \quad 1 \quad 2 \quad 3 \quad 4$$

Summary

Relations
1. A **relation** is any set of ordered pairs.
2. A **relation** may be represented by:
a. set of ordered pairs
b. table of values
c. graph
d. rule
e. an equation
3. The **domain** of a relation is the set of values that can be used for the **independent variable.**
4. The **range** of a relation is the set of values that represent the **dependent variable.**

Example 9
Each of the following represents the same relation.

   1. $((1, 3), (2, 5), (3, 7), (4, 9), (5, 11))$

   2. Table of values:

x	1	2	3	4	5
y	3	5	7	9	11

   3. Graph:

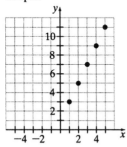

   4. Rule: For each natural number from 1 to 5 inclusive, double it and add 1 to obtain the corresponding y-value.

   5. Equation: $y = 2x + 1$ for $1 \leq x \leq 5, x \in N$.

The domain of this relation is the set $\{1, 2, 3, 4, 5\}$ and the range is the set $\{3, 5, 7, 9, 11\}$

Example 10
Postage stamps currently cost $.32. The cost C of n postage stamps is given by the rule $C = 0.32n$. This is an example of a relation. What is its domain and range?

Solution
The domain is the set of possible input values of this relation. Since n represents the number of stamps that can be purchased, $n \geq 0, n \in W$, (whole numbers). We call n the **independent variable** since it can be any whole number we choose. Once n is chosen, however, then the cost of n stamps, C, is uniquely determined. We say the variable C is the **dependent variable** since C depends upon the number of stamps purchased.

Summary

Functions

A **function** is a correspondence between a first set of elements, the domain, and a second set of elements, the range, such that each element of the domain corresponds to exactly one element in the range.

Example 11
Determine whether the figure represents a function or does not represent a function:

a.

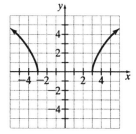

b.

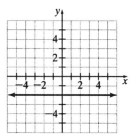

c.

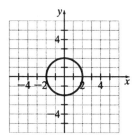

Solution
    a is a function, **b** is a function, **c** not a function

Example 12

    a. State the domain and range of the function graphed below.

    b. Use the graph to determine $f(0), f(-2)$ and $f(2)$.

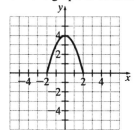

Solution

  a. The domain is $\{x|-2 \le x \le 2\}$. The range is $\{y|0 \le y \le 4\}$.

  b. From the graph, $f(0)$ is the *y*-value when *x* is zero. So, $f(0) = 4$, $f(2) = f(-2) = 0$.

Example 13
Evaluate the function indicated at $x = -2$:

  a. $f(x) = \sqrt[3]{-x^2 + 4x - 1}$ **b.** $g(x) = |2x - 9|$ **c.** $h(x) = -2x^2 + 3x$

Solution

  a. $f(x) = \sqrt[3]{-x^2 + 4x - 1}$
  $$f(-2) = \sqrt[3]{-(-2)^2 + 4(-2) - 1}$$
  $$= \sqrt[3]{-(4) + (-8) - 1}$$
  $$= \sqrt[3]{-13}$$

  b. $g(x) = |2x - 9|$
  $$g(-2) = |2(-2) - 9|$$
  $$= |-4 - 9|$$
  $$= |-13|$$
  $$= 13$$

  c. $h(x) = -2x^2 + 3x$
  $$h(-2) = -2(-2)^2 + 3(-2)$$
  $$= -2(4) + (-6)$$
  $$= -14$$

Example 14
A box in the shape of a rectangular solid is to have a length which is twice the width. The height of the box is to be 2 inches. Find a function, $V(x)$, which expresses the volume of the box as a function of its width, *x*, in cubic inches. Then find (if possible), $V(4)$, $V(3)$, $V(0)$ and $V(-2)$ and interpret the results.

Solution
The volume of the box can be found using the formula: $V = l \cdot w \cdot h$.
If $w = x$, then $l = 2x$. We are given that $h = 2$. Thus, $V(x) = (2x) \cdot x \cdot 2 = 4x^2$. $V(4)$ represents the volume of the box if the width is 4 inches.
$V(4) = 4(4)^2 = 4(16) = 64$ cubic inches
$V(3) = 4(3)^2 = 36$ cubic inches
$V(0) = 4(0)^2 = 0$
(The volume of the box is zero cubic inches if the width of the box is zero).
Finally, $V(-2)$ represents the volume of the box when the width is –2 inches.
Since a dimension cannot be negative, $V(-2)$ does not make sense. The domain of $V(x)$ is $\{x|x \ge 0\}$.

Exercise Set 8.2

1. Describe the set $\{5, 6, 7, 8, \ldots\}$.

2. Illustrate the set $\{x | x > 5 \text{ and } x \in N\}$ on a number line.

3. Illustrate the set $\{x | x > 5\}$ on a number line.

4. Express ←┼┼•┼•┼•┼•┼•┼┼┼┼→ in set builder notation.
   $$-4 \quad -3 \quad -2 \quad -1 \quad 0 \quad 1 \quad 2 \quad 3 \quad 4$$

Determine which of the relations are functions. Give the domain and range of each relation or function. For exercises 8–14 state the domain and range based upon the graph. When arrows are used, the graph extends in that direction.

5. $\{(1, 1), (2, 4), (3, 9), (4, 16)\}$

6. $\{(-1, 1), (-1, 4), (4, 8), (6, 9), (7, 10)\}$

7. $\{(-2, 3), (-1, 3), (0, 3), (4, 3), (8, 3)\}$

8.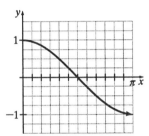

9.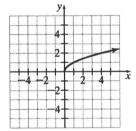

10.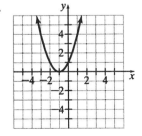

11.

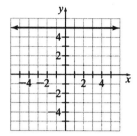

12.

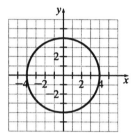

13.

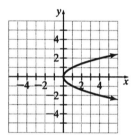

14.

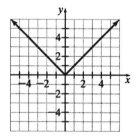

Evaluate the functions at the values indicated.

15. $f(x) = -2x^2 + 3x$ **a.** $f(-4)$ **b.** $f(2)$

16. $h(x) = 5 - 7x$ **a.** $h(0)$ **b.** $h(5)$

17. $g(x) = -\dfrac{3}{4}x + \dfrac{5}{6}$ **a.** $g(-4)$ **b.** $g(8)$

18. $f(x) = -x^3 - x^2 + 6x - 5$ **a.** $f(-1)$ **b.** $f(-2)$

19. $g(x) = \sqrt{-3x^2 + 5}$ **a.** $g(-3)$ **b.** $g(0)$

20. $h(x) = \dfrac{3x^2 + 4x}{-2x + 5}$ **a.** $h(4)$ **b.** $h(-3)$

21. $g(x) = |-3x + 7|$ **a.** $g(-3)$ **b.** $g(9)$

22. $f(x) = \dfrac{1}{2}x^3 + 3x$ **a.** $f\left(\dfrac{1}{2}\right)$ **b.** $f\left(-\dfrac{2}{3}\right)$

23. The height of a rock thrown from ground level at a speed of 80 feet per second is given by the function $h(t) = -16t^2 + 80t$. Find and interpret $h(0)$, $h(2.5)$ and $h(5)$.

24. A peach orchard near Macon, Georgia has 20 peach trees per acre. Each tree currently produces, on the average, 300 peaches. For each additional tree planted beyond 20, the average number of peaches per tree will be reduced by 10.
The function $Y(x) = (20 + x) \cdot (300 - 10x)$ gives the total yield of peaches for the peach orchard. Find and interpret $Y(0)$, $Y(5)$, and give the largest possible value of x.

Answers to Exercise Set 8.2

1. The set of natural numbers greater than 4.

2.

3.

4. $\{x| -3 \le x \le 1 \text{ and } x \in I\}$

5. A function; Domain = {1, 2, 3, 4}; Range = {1, 4, 9, 16}

6. Not a function; Domain = {−1, 4, 6, 7}; Range = {1, 4, 8, 9, 10}

7. A function; Domain = {−2, −1, 0, 4, 8}; Range = {3}

8. A function; Domain = $\{x|x \text{ is a real number}\}$; Range = $\{y|-1 \le y \le 1\}$

9. A function; Domain = $\{x|x \ge 0\}$; Range = $\{y|y \ge 0\}$

10. A function; Domain = $\{x|x \text{ is a real number}\}$; Range = $\{y|y \ge 0\}$

11. A function; Domain = $\{x|x \text{ is a real number}\}$; Range = {5}

12. Not a function; Domain = $\{x|-4 \le x \le 4\}$; Range = $\{y|-4 \le y \le 4\}$

13. Not a function; Domain $= \{x | x \geq 0\}$; Range $= \{y | y \text{ is real number}\}$

14. A function; Domain $= \{x | x \text{ is a real number}\}$; Range $= \{y | y \geq 0\}$

15. **a.** $f(-4) = -44$ **b.** $f(2) = -2$

16. **a.** $h(0) = 5$ **b.** $h(5) = -30$

17. **a.** $g(-4) = \dfrac{23}{6}$ **b.** $g(8) = -\dfrac{31}{6}$

18. **a.** $f(-1) = -11$ **b.** $f(-2) = -13$

19. **a.** $g(-3)$ is not a real number.   **b.** $g(0) = \sqrt{5}$

20. **a.** $h(4) = -\dfrac{64}{3}$ **b.** $h(-3) = \dfrac{15}{11}$

21. **a.** $g(-3) = 16$ **b.** $g(9) = 20$

22. **a.** $f\left(\dfrac{1}{2}\right) = \dfrac{25}{16}$ **b.** $f\left(-\dfrac{2}{3}\right) = -\dfrac{58}{27}$

23. $h(0) = 0$; the height of the ball initially is zero feet.
    $h(2.5) = 100$; the ball is 100 feet high after 2.5 seconds.
    $h(5) = 0$; the ball hits the ground after 5 seconds.

24. $Y(0) = 6000$; this is the yield of peaches if no additional trees are planted. $Y(5) = 6,250$; this is the yield if 5 additional trees are planted. The largest value of x is 30 since the term $(300 - 10x)$ must be nonnegative.

8.3   Linear Functions

Summary

Linear Equations

1. The **standard form of a linear equation** is given by: $ax + by = c$, where a and b are not both equal to zero.

2. A linear function is a function of the form $f(x) = ax + b$.

3. A **graph** of an equation is an illustration of the set of points whose coordinates satisfy the equation.

4. The graphs of all linear equations are **lines**. For that reason, it is only required to find and plot **two ordered pairs** that satisfy the equation. It is a good idea to find and plot a **third** point as a check.

Example 1
Graph $f(x) = 2x + 5$

Solution
Pick three arbitrary values of x, substitute into the equation and determine the three corresponding y-values:

	$y = f(x) = 2x + 5$	ordered pair
$x = 0$	$y = 2(0) + 5 = 5$	$(0, 5)$
$x = 2$	$y = 2(2) + 5 = 9$	$(2, 9)$
$x = -4$	$y = 2(-4) + 5 = -3$	$(-4, -3)$

Plot the three ordered pairs on a grid and connect them with a straight line:

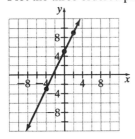

Example 2
Graph $-4x + 2y = -8$.

Solution
First, solve the equation for y:
$$-4x + 2y = -8$$
$$2y = 4x - 8$$
$$y = 2x - 4$$
Choose three arbitrary values of x, substitute these values into the equation and solve the resulting equation for y. It is a good idea to always choose at least one negative value for x.

	$y = 2x - 4$	ordered pair
$x = 0$	$y = 2(0) - 4 = -4$	$(0, -4)$
$x = -2$	$y = 2(-2) - 4 = -8$	$(-2, -8)$
$x = 3$	$y = 2(3) - 4 = 2$	$(3, 2)$

Plot the three ordered pairs and connect the points with a straight line:

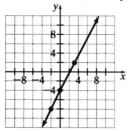

Example 3
Graph $\dfrac{1}{6}x - \dfrac{3}{4}y = \dfrac{3}{2}$

Solution
This equation, like the two previous examples, is a linear equation since it is of the form $ax + by = c$. Therefore, the graph of this equation is also a line.

First, multiply each side of the equation by the LCD of 12 to eliminate all fractions:
$$12\left(\frac{1}{6}x - \frac{3}{4}y\right) = 12 \cdot \frac{3}{2}$$
$$12\left(\frac{1}{6}x\right) - 12\left(\frac{3}{4}y\right) = \frac{12}{1} \cdot \frac{3}{2}$$
$$2x - 9y = 18$$

Now, solve the equation $2x - 9y = 18$ for y:
$$2x - 9y = 18$$
$$-9y = -2x + 18$$
$$y = \frac{2}{9}x - 2$$

Choose three arbitrary x-values and substitute these values into the equation to find three corresponding values of y. Choose x values which are multiplies of 9 to make the calculations easier.

$x = 9$ $y = \dfrac{2}{9}(9) - 2 = 0$ $(9, 0)$

$x = -9$ $y = \dfrac{2}{9}(-9) - 2 = -4$ $(-9, -4)$

$x = 18$ $y = \dfrac{2}{9}(18) - 2 = 2$ $(18, 2)$

Plot the three points and connect them with a straight line:

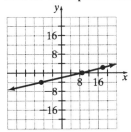

Summary

X- and *Y*-intercepts

1. The *x*-**intercept** is the point where the graph of an equation crosses the *x*-axis. It is found by letting $f(x) = 0$ and solving the equation for *x*.

2. The *y*-**intercept** is the point where the graph of an equation crosses the *y*-axis. It is found by letting $x = 0$ and solving the equation for $f(x)$.

3. The intercepts are convenient points to use when graphing linear equations.

Example 4

Graph $f(x) = -\dfrac{5}{4}x + 5$ by plotting intercepts.

Solution
Find the *y*-intercept by letting $x = 0$:

$f(x) = -\dfrac{5}{4}x + 5$

$f(0) = -\dfrac{5}{4}(0) + 5 = 5$

Thus, $(0, 5)$ is the *y*-intercept.

Find the *x*-intercept by letting $f(x) = 0$:

$$f(x) = -\frac{5}{4}x + 5$$

$$0 = -\frac{5}{4}x + 5$$

$$-5 = -\frac{5}{4}x$$

$$4 = x \quad \text{Multiply both sides by } -\frac{4}{5}.$$

Thus, $(4, 0)$ is the *x*-intercept.

Plot $(0, 5)$ and $(4, 0)$ and draw the graph:

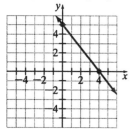

Summary

+--+
| **Special Case** |
| |
| The graph of $F(x) = b$ will be a horizontal line b units |
| above or below the *x*-axis, depending upon the sign of b. |
+--+

Example 5
Graph $f(x) = -4$.

Solution

For each *x*-value selected, $f(x)$ will **always be –4**. For instance, let $x = -2$, then $f(x) = -4$, $y = -4$. Let $x = 3$, then $f(x) = -4$. The graph will be a **horizontal** line 4 units **below** the *x*-axis.

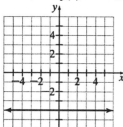

Example 6
A car's average speed is 55 miles per hour. Presently, the car is 30 miles from a certain city. Using the formula $d = rt$, (distance = rate × time), draw a graph of distance of the car (from the city) versus time (in hours).

Solution
Using $d = rt$, we have $D(t) = 55t + 30$ where 30 represents the starting point from the city. It is of the form $y = 55t + 30$. Thus, the graph of this equation will be a line.

Pick two arbitrary values of t, such as $t = 1$ and $t = 2$ and substitute the values into the equation to solve for d:

If $t = 1$, then $D(1) = 55(1) + 30 = 85$
If $t = 2$, then $D(2) = 55(2) + 30 = 140$

Plot $(1, 85)$ and $(2, 140)$ and draw a line connecting these points. Note: You will have to adjust the vertical scale on the y-axis to sketch a complete graph. Each tick mark on the vertical scale could represent 10 or 15 miles. Thus, you would need about 10 to 15 tick marks on the vertical scale to be able to plot the y-value of 140.

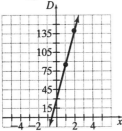

Exercise Set 8.3

Graph each function by plotting points.

1. $f(x) = -3$

2. $f(x) = -2x + 4$

3. $g(x) = \dfrac{3}{2}x + 3$

4. $g(x) = -\dfrac{2}{3}x + 1$

5. $p(x) = 4x - 3$

6. $d(x) = -\dfrac{2}{3}x + \dfrac{4}{3}$

7. $d(x) = \dfrac{10}{3}x - 5$

Graph using intercepts.

8. $f(x) = \dfrac{5}{2}x + 5$

9. $g(x) = \dfrac{1}{3}x + 2$

10. $h(x) = \dfrac{2}{7}x + 2$

11. A mechanic's bill is \$40.00 per hour labor in addition to \$150.00 for parts on a certain job. Draw a graph of the mechanic's bill (y) versus number of hours worked (x) from $0 \le x \le 10$.

12. The total cost to produce an item sold by a company consists of fixed costs of $80.00 plus a cost of $45.00 for each item produced.

   a. Draw a graph of the cost versus number of items produced from 0 to 15 items.

   b. Approximately how many items can be produced at a cost of $700.00?

   c. Find the cost of producing 9 items.

Answers to Exercise Set 8.3

1.

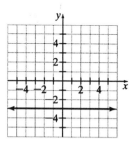

2.

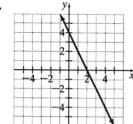

3.

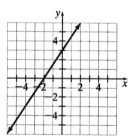

4.

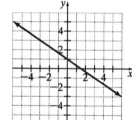

5.

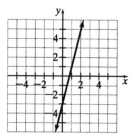

6.

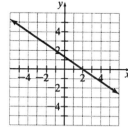

7.

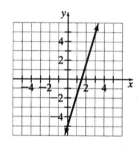

8.

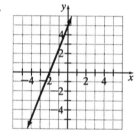

9.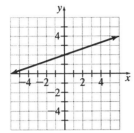

10.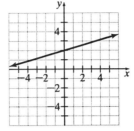

11.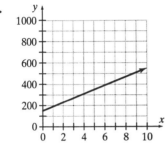

12. a.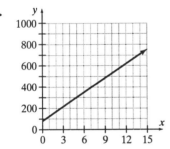

   b. 13 items

   c. Cost equals $485.00.

8.4   Slope, Modeling, and Linear Relationships

Summary

Slope

1. The **slope** of a line is a ratio of the vertical change to the horizontal change between **any two points** on the line.

2. The **slope** of the line through $P(x_1, y_1)$ and $Q(x_2, y_2)$ is given by
$$m = \frac{\text{change in } y}{\text{change in } x} = \frac{y_2 - y_1}{x_2 - y_1} = \frac{\Delta y}{\Delta x}$$
provided that $x_1 \neq x_2$.

Example 1
Find the slope of the line containing $P(-2, 5)$ and $Q(4, 7)$.

Solution

Use the formula $m = \dfrac{y_2 - y_1}{x_2 - x_1}$.

$$m = \frac{7 - 5}{4 - (-2)} = \frac{2}{6} = \frac{1}{3}$$

Summary

Positive and Negative Slopes Zero and Undefined Slopes

1. The slope of **any horizontal line** is zero.

2. The slope of **any vertical line** is undefined.

3. A line that rises going from left to right has a **positive slope.**

4. A line that falls going from left to right has a **negative slope.**

Example 2
The table below give an individual's salary in a given year:

Year:	Salary:
1973	$9,000
1983	$19,000
1987	$36,000
1990	$43,000
1995	$50,000

The table can be thought of as ordered pairs in which the first coordinate is the year and the second coordinate is the salary.

Find the slope of the line segments between

    a. 1973 and 1983 and

    b. between 1987 and 1995.

Solution

    a. $m = \dfrac{\text{vertical change}}{\text{horizontal change}} = \dfrac{19,000 - 9000}{1983 - 1973} = \dfrac{10,000}{10} = 1000$
Interpretation: The slope of 1000 represents the yearly average change in annual salary over the ten year period from 1973 to 1983.

    b. $m = \dfrac{\text{vertical change}}{\text{horizontal change}} = \dfrac{50,000 - 36,000}{1995 - 1987} = \dfrac{14,000}{8} = 1750$
This can be interpreted as an increase of $1750 in income per year over the 8-year period from 1987 to 1995.

Summary

> **Slope-Intercept Form of a Linear Equation**
>
> $y = mx + b$ where m is the **slope** and b is the **y-intercept**.

Example 3
Write the equation $-2x + 4y = 8$ in slope-intercept form. State the slope and the y-intercept.

Solution
First solve for y: $-2x + 4y = 8$

$$4y = 2x + 8$$

$$y = \frac{1}{2}x + 2$$

To graph the line, start at the point $(0, 2)$, (y-intercept) and move vertically upward 1 unit and horizontally to the right 2 units to obtain a second point, $(2, 3)$. Repeat this procedure to obtain a third point $(4, 4)$. Plot all three points and connect them with a straight line.

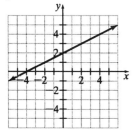

Summary

Point-Slope Form of a Linear Equation

$y - y_1 = m(x - x_1)$, where m is the slope and (x_1, y_1) is a specific point on the line and (x, y) is another arbitrary point on the line.

Example 4
Find the equation of a line, in slope-intercept form, whose slope is -4 which passes through $P(-2, 5)$.

Solution
Let $Q(x, y)$ be **any arbitrary** point on the line and $P(-2, 5)$ be a specific point. Use the point-slope form for the equation of a line:

$y - y_1 = m(x - x_1)$
$y - 5 = -4(x - (-2))$
$y - 5 = -4(x + 2)$
$y - 5 = -4x - 8$
$\quad\quad y = -4x - 3$

Example 5
Find the slope intercept form of the equation of a line which passes through $P(-2, 7)$ and $Q(4, -9)$.

Solution

First, find the slope of the line using the slope formula: $m = \dfrac{y_2 - y_1}{x_2 - x_1} = \dfrac{-9 - 7}{4 - (-2)} = \dfrac{-16}{6} = -\dfrac{8}{3}$

Now, use the point slope formula and choose either P or Q as (x_1, y_1). P will be used in our example:

$$y - y_1 = m(x - x_1)$$
$$y - 7 = -\frac{8}{3}(x - (-2))$$
$$y - 7 = -\frac{8}{3}(x + 2)$$
$$y - 7 = -\frac{8}{3}x + \frac{-16}{3}$$
$$y = -\frac{8}{3}x - \frac{16}{3} + \frac{21}{3}$$
$$y = -\frac{8}{3}x + \frac{5}{3}$$

Summary

Parallel and Perpendicular Lines

1. Two lines are **parallel** if and only if their slopes are the same.

2. Two lines are **perpendicular** if and only if the product of their slopes is negative one. This means that their slopes are negative reciprocals of each other.

Example 6
Determine if the two lines are parallel, perpendicular or neither.

   a. L_1: (3, 5), (4, 7); L_2: (8, 7), (10, 6)

   b. L_1: (7, 10), (11, 13); L_2: (8, 11), (12, 14)

Solution

   a.   The slope of L_1 is $\dfrac{7 - 5}{4 - 3} = 2$. The slope of L_2 is $\dfrac{6 - 7}{10 - 8} = -\dfrac{1}{2}$. Since $2 \cdot -\dfrac{1}{2} = -1$, the lines are perpendicular.

   b.   The slope of L_1 is $m = \dfrac{13 - 10}{11 - 7} = \dfrac{3}{4}$. The slope of L_2 is $m = \dfrac{14 - 11}{12 - 8} = \dfrac{3}{4}$. Since the slopes are equal, the lines are parallel.

Example 7
Determine if the graphs of the equations are parallel or perpendicular lines.
$-2x + 3y = 6$
$6x - 9y = 8$

Solution
Solve each equation for y and note the coefficient of the x term. This coefficient will be the slope of the line.
$-2x + 3y = 6$
$$3y = 2x + 6$$
$$y = \frac{2}{3}x + 2$$

Thus, the slope of the first line is $\frac{2}{3}$.

$6x - 9y = 8$
$$-9y = -6x + 8$$
$$y = \frac{-6x}{-9} + \frac{8}{-9}$$
$$y = \frac{2}{3}x - \frac{8}{9}$$

Thus, the slope of the second line is $\frac{2}{3}$.

Since the slopes are the same, the lines are parallel.

Example 8
Consider the equation $3x - 5y = 15$. Determine the equation of the line which has a y-intercept of 6 and is

   a. parallel to the given line and

   b. perpendicular to the given line.

Solution
Solve $3x - 5y = 15$ for y to find the slope.
$3x - 5y = 15$
$$-5y = -3x + 15$$
$$y = \frac{3}{5}x - 3$$

Thus, the slope of the line is $\frac{3}{5}$.

a. The slope of the parallel line is also $\dfrac{3}{5}$.

use the point slope formula with the given point $P(0, 6)$, (y-intercept) and the slope of $\dfrac{3}{5}$.

$$y - y_1 = m(x - x_1)$$
$$y - 6 = \frac{3}{5}(x - 0)$$
$$y - 6 = \frac{3}{5}x$$
$$y = \frac{3}{5}x + 6$$

b. Again, use the point-slope formula with the given point $(0, 6)$ but this time use a slope of $-\dfrac{5}{3}$, since $-\dfrac{5}{3}$

is the negative reciprocal of $\dfrac{3}{5}$.

$$y - y_1 = m(x - x_1)$$
$$y - 6 = -\frac{5}{3}(x - 0)$$
$$y - 6 = -\frac{5}{3}x$$
$$y = -\frac{5}{3}x + 6$$

Exercise Set 8.4

Find the slope of the line through the given points. If the slope is undefined, then so state.

1. $(3, 5)$ and $(9, 7)$ **2.** $(-3, 2)$ and $(4, 16)$

3. $(6, -2)$ and $(-1, -2)$ **4.** $(-3, 5)$ and $(-3, 10)$

5. $(-5, -11)$ and $(4, -3)$

Solve for x if the line through the given points is to have the given slope.

6. $(1, 0), (5, x)$; slope is -3 **7.** $(-4, -1), (x, 2)$; slope $= -\dfrac{3}{5}$

8. Find the slope of the line in each figure below:

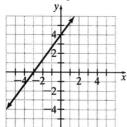

Write each equation in slope-intercept form and indicate the slope and the y-intercept. Then, use the y-intercept and slope to draw the graph of the linear equation.

9. $3x - y = -2$ **10.** $60x = 30y + 60$ **11.** $-2x - 9y = 18$

Two points on L_1 and two points on L_2 are given. Determine if L_1 is parallel to, perpendicular to, or neither parallel nor perpendicular to L_2.

12. L_1: $(9, -5)$ and $(4, 6)$; L_2: $(3, -2)$ and $(-2, -5)$

13. L_1: $(4, 7)$ and $(6, 12)$; L_2: $(-3, -3)$ and $(-1, 2)$

14. L_1: $(7, 10)$ and $(11, 13)$; L_2: $(8, 11)$ and $(12, 14)$

Determine if the given lines are parallel, perpendicular or neither:

15. $-2x + y = 7$ and $-4x + 2y = 10$

16. $y = -\dfrac{3}{4}x + 7$ and $y = \dfrac{4}{3}x - 5$

17. $x - 3y = -9$ and $y = 3x + 6$

Write the equation of the line in point-slope form given the following properties of the line:

18. line has slope -3 and passes through $(-2, 4)$

19. line passes through $(-2, 5)$ and $(3, -7)$

20. line passes through $(-1, 8)$ and $(7, 16)$

21. through $(2, 5)$ and parallel to the line with x-intercept and y-intercept 3.

22. Through $(0, 0)$ and perpendicular to the line whose equation is $\dfrac{1}{5}x - \dfrac{2}{7}y = 1$

23. An office machine was purchased in 1980 at a cost of $100,000. In the year 2000, the machine will have a market value of 0 dollars. Assuming that the relationship between the value of the machine (V) and the time (t) in years is linear, find an equation of the line depicting value versus time. Use this equation to find the value of the machine in 1995.

Answers to Exercise Set 8.4

1. $\dfrac{1}{3}$

2. 2

3. 0

4. Slope is undefined.

5. $\dfrac{8}{9}$

6. $x = -12$

7. $x = -9$

8. $m = \dfrac{4}{3}$

9. $y = 3x + 2$, $m = 3$, y-intercept $(0, 2)$

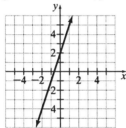

10. $y = 2x - 2$; $m = 2$, y-intercept $(0, -2)$

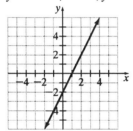

11. $y = -\dfrac{2}{9}x - 2$; $m = -\dfrac{2}{9}$; $b = -2$

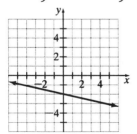

12. Neither

13. Parallel

14. Parallel

15. Parallel

16. Perpendicular

17. Neither

18. $y = -3x + 2$

19. $y = -\dfrac{12}{5}x + \dfrac{1}{5}$

20. $y = x + 9$

21. $y = -x + 7$

22. $y = -\dfrac{10}{7}x$

23. $V = -5000t + 100,000$

At $t = 15$, then $V = 25,000$

8.5 The Algebra of Functions

Summary

Algebra of Functions

If $f(x)$ represents one function and $g(x)$ represents a second function, then the following operations on functions may be performed:

Sum of functions: $(f + g)(x) = f(x) + g(x)$

Difference of functions: $(f - g)(x) = f(x) - g(x)$

Product of functions: $(f \cdot g)(x) = f(x) \cdot g(x)$

Quotient of functions: $\left(\dfrac{f}{g}\right)(x) = \dfrac{f(x)}{g(x)}$

$$g(x) \neq 0$$

Example 1

Let $f(x) = 2x^2 + 3x - 1$; $g(x) = 3x - 2$. Find the following:

a. $(f + g)(x)$

b. $(f - g)(x)$

c. $(g - f)(x)$

d. $(f \cdot g)(x)$

e. $(f + g)(2)$

f. $\left(\dfrac{f}{g}\right)\left(\dfrac{2}{3}\right)$

Solution

a.
$$\begin{aligned}
(f + g)(x) &= f(x) + g(x) \\
&= 2x^2 + 3x - 1 + (3x - 2) \\
&= 2x^2 + 6x - 3
\end{aligned}$$

b.
$$\begin{aligned}
(f - g)(x) &= f(x) - g(x) \\
&= 2x^2 + 3x - 1 - (3x - 2) \\
&= 2x^2 + 3x - 1 - 3x + 2 \\
&= 2x^2 + 1
\end{aligned}$$

c. $(g - f)(x) = g(x) - f(x)$

$$= 3x - 2 - (2x^2 + 3x - 1)$$
$$= 3x - 2 - 2x^2 - 3x + 1$$
$$= -2x^2 - 1$$

d. $(f \cdot g)(x) = f(x) \cdot g(x)$

$$= (2x^2 + 3x - 1)(3x - 2)$$
$$= 6x^3 - 4x^2 + 9x^2 - 6x - 3x + 2$$
$$= 6x^3 + 5x^2 - 9x + 2$$

e. $(f + g)(2) = f(2) + g(2)$

$$= 2(2)^2 + 3(2) - 1 + (3(2) - 2)$$
$$= 8 + 6 - 1 + 4$$
$$= 17$$

f. $\left(\dfrac{f}{g}\right)\left(\dfrac{2}{3}\right) = \dfrac{f\left(\frac{2}{3}\right)}{g\left(\frac{2}{3}\right)} = \dfrac{2\left(\frac{2}{3}\right)^2 + 3\left(\frac{2}{3}\right) - 1}{3\left(\frac{2}{3}\right) - 2} = \dfrac{\frac{17}{9}}{0}$

Since division by 0 is not permitted, the expression $\left(\dfrac{f}{g}\right)\left(\dfrac{2}{3}\right)$ is not defined.

Example 2
Using the graphs of $f(x)$ and $g(x)$, find

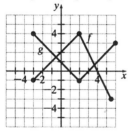

a. $(f + g)(1)$ **b.** $(f - g)(5)$ **c.** $(f \cdot g)(5)$

Solution

a. $(f + g)(1) = f(1) + g(1)$ **b.** $(f - g)(5) = f(5) - g(5)$ **c.** $(f \cdot g)(5) = f(5) \cdot g(5)$
$$= 3 + 0 \qquad\qquad\qquad\qquad = (-2) - 2 \qquad\qquad\qquad\qquad = (-2)(2)$$
$$= 3 \qquad\qquad\qquad\qquad\quad = -4 \qquad\qquad\qquad\qquad\qquad = -4$$

Exercise Set 8.5

For exercises 1–3, use $f(x)$ and $g(x)$ to find

   a. $(f+g)(x)$ **b.** $(f-g)(x)$

   c. $(f \cdot g)(x)$ **d.** $\left(\dfrac{f}{g}\right)(x)$

1. $f(x) = 3x$ and $g(x) = x^2$

2. $f(x) = 5x - 2$ and $g(x) = x^2 - 3x + 1$

3. $f(x) = 2x^2 - x + 5$ and $g(x) = 3x + 7$

Given $f(x) = x^2 - 4x + 5$ and $g(x) = 2x - 3$, find

   4. $(f \cdot g)(3)$ **5.** $(f-g)(2)$

Given $f(x) = 2x^2 - 3x + 2$ and $g(x) = 5x - 2$, find

   6. $(f+g)(-1)$ **7.** $\left(\dfrac{f}{g}\right)(1)$

Answers to Exercise Set 8.5

1. a. $3x + x^2$ **b.** $3x - x^2$ **c.** $3x^3$ **d.** $\dfrac{3}{x}$

2. a. $x^2 + 2x - 1$ **b.** $-x^2 + 8x - 3$ **c.** $5x^3 - 17x^2 + 11x - 2$ **d.** $\dfrac{5x - 2}{x^2 - 3x + 1}$

3. a. $2x^2 + 2x + 12$ **b.** $2x^2 - 4x - 2$ **c.** $6x^3 + 11x^2 + 8x + 35$ **d.** $\dfrac{2x^2 - x + 5}{3x + 7}$

4. 6   **5.** 0   **6.** 0   **7.** $\dfrac{1}{3}$

Chapter 8 Practice Test

   1. Are the lines $2x + 3y = 5$ and $2x - 3y = 7$ parallel?

   2. Find the slope and y-intercept of $-3x + 6y = 18$.

   3. Write the equation of a line passing through the points $(-2, 3)$ and $(5, 10)$.

   4. Write the equation of a line through $(0, 3)$ and perpendicular to the line $-3x + 5y = 10$.

5. Graph $f(x) = -\dfrac{3}{5}x + 1$.

6. Graph $g(x) = 2x + 4$, using x- and y-intercepts.

7. The distance that an automobile is from a certain city is estimated by the function $d(t) = 100 + 55t$ where t is in hours and d is in miles.

    a.   Graph $d(t)$ for $0 \le t \le 10$.

    b.   Find $d(0)$, $d(5)$ and interpret the results.

    c.   For $0 \le t \le 10$, find the range of this function.

8. For $f(x) = -3x^2 + 4x - 2$, find

    a. $f(-2)$ **b.** $f\left(\dfrac{2}{3}\right)$

9. Graph $f(x) = 5$.

10. Graph $g(x) = -3x + 2$.

For problems 11–12, find the domain and range of each relation and determine if the relation is a function.

11. $\{(3, 4), (-2, 6), (5, 8), (-2, 5)\}$

12.

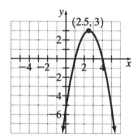

13. $f(x) = \sqrt{x^2 + 16}$, find

    a. $f(2)$ **b.** $f(-3)$ **c.** $f(0)$

Answers to Chapter 8 Practice Test

1. No

2. Slope is $\dfrac{1}{2}$; y-intercept $(0, 3)$

3. $y = x + 5$

4. $y = -\dfrac{5}{3}x + 3$

5.

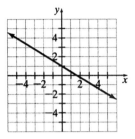

6.

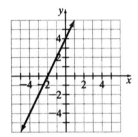

7. a.

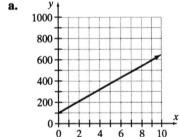

    b. $d(0) = 100$; $d(5) = 375$
       Interpretation: $d(0) = 100$ means that after $t = 0$ hours, the car is 100 miles from the city.
       $d(5) = 375$ means that after 5 hours, the car is 375 miles from this certain city.

    c. The range is $\{y | 0 \le y \le 650\}$.

8. a. $f(-2) = -22$ **b.** $f\left(\dfrac{2}{3}\right) = -\dfrac{2}{3}$

9.

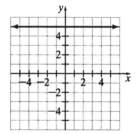

10.

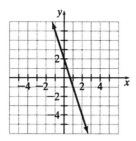

11. Domain = $\{-2, 3, 5\}$; range = $\{4, 5, 6, 8\}$; not a function

12. Domain = $\{x | x \text{ is a real number}\}$; range = $\{y | y \leq 3\}$, a function

13. a. $f(2) = \sqrt{20} = 2\sqrt{5}$ **b.** $f(-3) = 5$ **c.** $f(0) = 4$

Chapter 9

9.1 Solving Systems of Equations Graphically

Summary

A **system of linear equations** or **simultaneous linear equations** consist of two or more linear equations. The **solution to a system of equations** is the ordered pair or pairs that will satisfy all equations.

Example 1
Determine which of the following ordered pairs satisfy the system of equations.
$$3x + 4y = 18$$
$$2x - y = 1$$

   a. $(6, 0)$ **b.** $(3, 5)$ **c.** $(2, 3)$

Solution

  a. Substitute 6 for x and 0 for y in both equations.

$$3x + 4y = 18 \qquad\qquad 2x - y = 1$$
$$3(6) + 4(0) = 18 \qquad\qquad 2(6) - 0 = 1$$
$$18 + 0 = 18 \qquad\qquad 12 - 0 = 1$$
$$18 = 18 \text{ True} \qquad\qquad 12 = 1 \text{ False}$$

     Since $(6, 0)$ does not satisfy both equations, it is not a solution to the system of equations.

  b. Substitute 3 for x and 5 for y in both equations.

$$3x + 4y = 18 \qquad\qquad 2x - y = 1$$
$$3(3) + 4(5) = 18 \qquad\qquad 2(3) - 5 = 1$$
$$9 + 20 = 18 \qquad\qquad 6 - 5 = 1$$
$$29 = 18 \text{ False} \qquad\qquad 1 = 1 \text{ True}$$

     Since $(3, 5)$ does not satisfy both equations, it is not a solution of the system of equations.

  c. Substitute 2 for x and 3 for y in both equations.

$$3x + 4y = 18 \qquad\qquad 2x - y = 1$$
$$3(2) + 4(3) = 18 \qquad\qquad 2(2) - 3 = 1$$
$$6 + 12 = 18 \qquad\qquad 4 - 3 = 1$$
$$18 = 18 \text{ True} \qquad\qquad 1 = 1 \text{ True}$$

     Since $(2, 3)$ satisfies both equations, it is a solution to the system of equations.

Summary

> The **solution to a system of linear equations** is the ordered pair (or pairs) common to all lines in the system when the lines are graphed.
>
> If one point is common to both equations, the system of linear equations is **consistent**. The linear equations in this system have different slopes.
>
> If no points are common to both equations, the system of linear equations is **inconsistent.** The linear equations have the same slope but different y-intercepts.
>
> If an infinite number of points are common to both equations, the system of linear equations is **dependent.** The linear equations have the same slope and the same y-intercept.

Example 2
Determine if the system is consistent, inconsistent, or dependent.
$$2x - 3y = 8$$
$$-4x + 6y = -16$$

Solution
Write each in slope-intercept form and then compare the slopes and y-intercepts.

$$2x - 3y = 8 \qquad\qquad -4x + 6y = -16$$
$$-3y = -2x + 8 \qquad\qquad 6y = 4x - 16$$
$$y = \frac{2}{3}x - \frac{8}{3} \qquad\qquad y = \frac{2}{3}x - \frac{8}{3}$$

Since the equations have the same slopes and the same y-intercept, the system is dependent.

Summary

> **To obtain the solution to a system of equations graphically,** graph each equation and determine the point or points of intersection.

Example 3
Solve the following systems of equations graphically.

    a. $3x - 2y = 6$ **b.** $3x - y = 9$
        $3x + 2y = 6$ $6x - 2y = 12$

Solution

a. Solve each equation for y and graph using slope and y-intercept.

$$3x - 2y = 6 \qquad\qquad\qquad 3x + 2y = 6$$
$$-2y = -3x + 6 \qquad\qquad\qquad 2y = -3x + 6$$
$$y = \frac{3}{2}x - 3 \qquad\qquad\qquad y = -\frac{3}{2}x + 3$$
$$\text{(line 1)} \qquad\qquad\qquad\qquad \text{(line 2)}$$

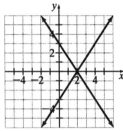

Solution is $(2, 0)$.

Check: Substitute 2 for x and 0 for y in both equations.

$$3x - 2y = 6 \qquad\qquad\qquad 3x + 2y = 6$$
$$3(2) - 2(0) = 6 \qquad\qquad\qquad 3(2) + 2(0) = 6$$
$$6 - 0 = 6 \qquad\qquad\qquad\qquad 6 + 0 = 6$$
$$6 = 6 \text{ True} \qquad\qquad\qquad\qquad 6 = 6 \text{ True}$$

b. Solve each equation for y and graph using the slope and y-intercept.

$$3x - y = 9 \qquad\qquad\qquad 6x - 2y = 12$$
$$-y = -3x + 9 \qquad\qquad\qquad -2y = -6x + 12$$
$$y = 3x - 9 \qquad\qquad\qquad y = 3x - 6$$
$$\text{(line 1)} \qquad\qquad\qquad\qquad \text{(line 2)}$$

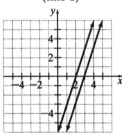

From the graph, the lines are parallel. The system is inconsistent and has no solution. Notice that the slope is the same but the two y-intercepts are different.

Exercise Set 9.1

Determine which, if any, of the following ordered pairs satisfy each system of linear equations.

1. $x + y = -3$
$x - y = 7$

    a. $(2, -5)$ **b.** $(2, 1)$ **c.** $(8, 1)$

2. $x + y = 8$
$2x + y = 12$

   a. $(6, 2)$ **b.** $(4, 4)$ **c.** $(3, 6)$

3. $x + y = 3$
$x - 2y = 12$

   a. $(1, 2)$ **b.** $(0, -6)$ **c.** $(6, -3)$

4. $2x + 4y = 9$
$x + y = 1$

   a. $(4, -3)$ **b.** $\left(0, \dfrac{9}{4}\right)$ **c.** $(8, -7)$

Express each equation in *y*-intercept form. Without graphing the equations, state whether the system has one solution, no solution, or an infinite number of solutions.

5. $4x - y = 6$
$2x + y = 3$

6. $x + y = 8$
$4x + 4y = 10$

7. $7x + 3y = 11$
$2x + 5y = 7$

8. $2x - 4y = 12$
$-x + 2y = -6$

Determine the solution to the following systems of equations graphically. If the system is dependent or inconsistent, so state.

9. $7x - y = 7$
$x - y = 1$

10. $2x + y = 8$
$2x - y = -4$

11. $x + y = 5$
$2x + 2y = 12$

12. $x - y = 2$
$3x - 3y = 6$

13. $x = 2y + 5$
$x = y + 1$

14. $2x - 3y = -11$
$x + 3y = 8$

Answers to Exercise Set 9.1

1. $(2, -5)$ **2.** $(4, 4)$ **3.** $(6, -3)$

4. none   **5.** one solution   **6.** no solution

7. one solution   **8.** infinite number of solutions   **9.** $(1, 0)$

10. $(1, 6)$ **11.** inconsistent   **12.** dependent

13. $(-3, -4)$ **14.** $(-1, 3)$

9.2 Solving Systems of Equations by Substitution

Summary

To Solve a System of Equations by Substitution

1. Solve for a variable in either equation. (If possible, solve for a variable with a numerical coefficient of 1 to avoid working with fractions.)

2. Substitute the expression found for the variable in step 1 into the other equation.

3. Solve the equation determined in step 2 to find the value of one variable.

4. Substitute the value of the variable determined in step 3 into the equation obtained in step 1 to find the other variable.

Example 1
Solve the following system of equations by substitution.
$$x + y = 3$$
$$2x - 3y = 1$$

Solution

(Step 1)   Solve for y in the first equation.
$$x + y = 3$$
$$y = 3 - x$$

(Step 2)   Substitute $3 - x$ for y in the second equation.
$$2x - 3(3 - x) = 1$$

(Step 3) $2x - 9 + 3x = 1$
$$5x - 9 = 1$$
$$5x = 10$$
$$x = 2$$

(Step 4)   Now, substitute $x = 2$ in the equation $y = 3 - x$.
$$y = 3 - x$$
$$y = 3 - 2$$
$$y = 1$$

So, the solution is the ordered pair (2, 1).

Example 2
Solve the following system of equations by substitution.
$$2x + 3y = 3$$
$$6x - 3y = 1$$

Solution

(Step 1)   Solve for x in the first equation. (Note, since there are no variables with a coefficient of 1, we cannot avoid fractions.)

$$2x + 3y = 3$$
$$2x = 3 - 3y$$
$$x = \frac{3 - 3y}{2}$$
$$x = \frac{3}{2} - \frac{3y}{2}$$

(Step 2)   Substitute $\frac{3}{2} - \frac{3y}{2}$ for x in the second equation.

$$6\left(\frac{3}{2} - \frac{3y}{2}\right) - 3y = 1$$

(Step 3) $9 - 9y - 3y = 1$
$$9 - 12y = 1$$
$$-12y = -8$$
$$y = \frac{-8}{-12} = \frac{2}{3}$$

(Step 4)   Now, substitute $y = \frac{2}{3}$ in the equation $x = \frac{3}{2} - \frac{3y}{2}$.

$$x = \frac{3}{2} - \frac{3y}{2}$$
$$x = \frac{3}{2} - \frac{3}{2}\left(\frac{2}{3}\right)$$
$$x = \frac{3}{2} - 1$$
$$x = \frac{1}{2}$$

So, the solution is $\left(\frac{1}{2}, \frac{2}{3}\right)$.

Example 3
Solve the following system of equations by substitution.
$y = 2x + 5$
$1 = 6x - 3y$

Solution

(Step 1)   The first equation is already solved for y. So, we will use $y = 2x + 5$.

(Step 2)   Substitute $2x + 5$ for y in the second equation.
$6x - 3(2x + 5) = 1$

(Step 3) $6x - 6x - 15 = 1$
$\qquad\qquad -15 = 1$
Since the statement $-15 = 1$ is false, the system has no solution. The graphs of the equations will be parallel lines and the system is inconsistent.

Example 4
Solve the following system of equations by substitution.
$x - 2y = 3$
$3x - 6y = 9$

Solution

(Step 1)   Solve for x in the first equation.
$x - 2y = 3$
$\quad x = 3 + 2y$

(Step 2)   Substitute $3 + 2y$ for x in the second equation.

(Step 3) $3(3 + 2y) - 6y = 9$
$\quad 9 + 6y - 6y = 9$
$\qquad\qquad 9 = 9$
Since $9 = 9$ is a true statement, this system has an infinite number of solutions. The graphs of the equations represent the same line when graphed and the system is dependent.

Exercise Set 9.2

Find the solution to each system of equations.

1. $y = x - 5$
    $x + y = 3$

2. $x = y + 1$
    $3x + 4y = 17$

3. $x + y = 9$
    $8x - y = -18$

4. $x + 8y = 6$
    $-2x - 4y = 3$

5. $5x = y + 8$
    $10x - 2y = 16$

6. $3x + 2y = 8$
    $y = -\dfrac{3}{2}x - 2$

7. $2x + y = 3$
    $4x + 3y = 10$

8. $x = 4y + 1$
    $3x - 8y = 8$

9. $2x + 2y = -12$
    $5y = 3x - 6$

10. $4x - 3y = 1$
$5x + 2y = 7$

Answers to Exercise Set 9.2

1. $(4, -1)$ **2.** $(3, 2)$ **3.** $(-1, 10)$

4. $\left(-4, \dfrac{5}{4}\right)$ **5.** infinite number of solutions **6.** no solution

7. $\left(-\dfrac{1}{2}, 4\right)$ **8.** $\left(6, \dfrac{5}{4}\right)$ **9.** $(-3, -3)$

10. $(1, 1)$

9.3 Solving Systems of Equations by the Addition Method

Summary

To Solve a System of Equations by the Addition (or Elimination) Method:

1. If necessary, rewrite each equation so that the terms containing the variables appear on the left side of the equal sign and any constants appear on the right side of the equal sign.

2. If necessary, multiply both equations by a constant(s) so that when the equations are added the resulting sum will contain only one variable.

3. Add the equations. This will result in a single equation containing only one variable.

4. Solve for the variable in the equation in step 3.

5. Substitute the value found in step 4 into either of the original equations. Solve that equation to find the value of the remaining variable.

6. Check the values obtained in all original equations.

Example 1
Solve the following equation using the addition method.
$3x - 2y = 10$
$3x + 2y = 14$

Solution
Steps 1 and 2 are unnecessary.

(Step 3) $3x - 2y = 10$
             $\underline{3x + 2y = 14}$
             $6x \quad\quad = 24$

(Step 4) $6x = 24$
             $x = \dfrac{24}{6}$
             $x = 4$

(Step 5)   Substitute $x = 4$ into the second equation.
             $3x + 2y = 14$
             $3(4) + 2y = 14$
             $12 + 2y = 14$
             $2y = 2$
             $y = \dfrac{2}{2} = 1$

(Step 6)   A check shows that $x = 4$ and $y = 1$ are the correct values.

Thus, the solution to the system of equations is (4, 1).

Example 2
Solve the following system of equations using the addition method.
$2x + 3y = 2$
$3x = 2y + 16$

Solution

(Step 1) $2x + 3y = 2$
             $\underline{3x - 2y = 16}$

(Step 2)   Multiply the first equation by 2 and the second equation by 3 so that upon addition the y variable will be eliminated.
             $2[2x + 3y = 2]$ gives $4x + 6y = 4$
             $3[3x - 2y = 16]$ gives $9x - 6y = 48$

(Step 3) $4x + 6y = \quad 4$
             $\underline{9x - 6y = 48}$
             $13x \quad\quad = 52$

(Step 4) $13x = 52$

$$x = \frac{52}{13} = 4$$

(Step 5)   Substitute $x = 4$ into the first equation.

$$2x + 3y = 2$$
$$2(4) + 3y = 2$$
$$8 + 3y = 2$$
$$3y = -6$$
$$y = \frac{-6}{3} = -2$$

(Step 6)   A check shows that $x = 4$ and $y = -2$ are the correct values.

Thus, the solution to the system of equations is $(4, -2)$.

Example 3
Solve the following system of equations using the addition method.
$$3x - 6y = 8$$
$$x - 2y = 2$$

Solution
Step 1 is unnecessary.

(Step 2)   Multiply the second equation by -3 so that upon addition the y variable will be eliminated.

$$3x - 6y = 8 \qquad\qquad 3x - 6y = 8$$
$$-3[x - 2y = 2] \text{ gives } -3x + 6y = -6$$

(Step 3)
$$\begin{array}{r} 3x - 6y = 8 \\ -3x + 6y = -6 \\ \hline 0 = 2 \end{array}$$

(Step 4)   Since $0 = 2$ is a false statement, the system has no solution. The graphs of the equations are parallel lines and the system is inconsistent.

Example 4
Solve the following system of equations using the addition method.
$$4x + 8y = 12$$
$$x = 3 - 2y$$

Solution

(Step 1) $4x + 8y = 12$
             $x + 2y = 3$

(Step 2)   Multiply the second equation by -4 to eliminate the x variable.

$$4x + 8y = 12 \qquad\qquad 4x + 8y = 12$$
$$-4[x + 2y = 3] \text{ gives } -4x - 8y = -12$$

(Step 3) $4x + 8y = 12$
                  $\underline{-4x - 8y = -12}$
                          $0 = 0$

(Step 4)   Since $0 = 0$ is a true statement, there is an infinite number of solutions to the system. The graphs of the equations will be the same line and the system is dependent.

Exercise Set 9.3

Solve the following systems of equations using the addition method.

1. $x + y = -1$
    $x - y = 3$

2. $2x - y = 3$
    $-2x + y = 2$

3. $x + 5y = 4$
    $-x - 5y = -4$

4. $3x - 2y = -11$
    $x = -1 - 2y$

5. $y = 1 - 2x$
    $6x + 5y = 13$

6. $x + 4y = 2$
    $x - 8y = -7$

7. $2x + 4y = 8$
    $3x + 6y = 6$

8. $7x - 5y = 19$
    $3x - 2y = 6$

9. $3x - 5y = 10$
    $7x - 3y = 14$

10. $5x + 7y = 1$
     $7x + 5y = -1$

Answers to Exercise Set 9.3

1. $(1, -2)$

2. no solution

3. infinite number of solutions

4. $(-3, 1)$

5. $(-2, 5)$

6. $\left(-1, \dfrac{3}{4}\right)$

7. no solution

8. $(-8, -15)$

9. $\left(\dfrac{20}{13}, -\dfrac{14}{13}\right)$

10. $\left(-\dfrac{1}{2}, \dfrac{1}{2}\right)$

9.4   Solving Systems of Linear Equations in Three Variables

Summary

A third-order system consists of three equations and three unknowns. The substitution method or addition method introduced in Section 4.1 can be used to solve these systems.

Example 1
Solve the following system using the substitution method.
$x + 2y + 3z = -2$
$\quad\quad y + z = -1$
$\quad\quad\quad z = -2$

Solution:
Substitute $z = -2$ into $y + z = -1$ and solve for y.
$\quad y + z = -1$
$y + (-2) = -1$
$\quad\quad y = 1$
Now, substitute $y = 1$ and $z = -2$ into $x + 2y + 3z = -2$ and solve for x.
$\quad x + 2y + 3z = -2$
$x + 2(1) + 3(-2) = -2$
$\quad\quad x + 2 - 6 = -2$
$\quad\quad\quad\quad x = 2$
So, the solution is the ordered triple (2, 1, –2).

Example 2
Solve the following system by the addition method.
$\quad x + 2y + z = 3 \quad$ (1)
$\quad x - 3y - z = 2 \quad$ (2)
$x - 2y - 2z = -2 \quad$ (3)

Solution
In using the addition method, we choose two equations and attempt to eliminate a variable. Then choose another two equations and attempt to eliminate the same variable. The result will be two equations with two variables which then can be solved.
Here, add equations (1) and (2) to eliminate z.
$x + 2y + z = 3$
$\underline{x - 3y - z = 2}$
$\quad 2x - y = 5$ (4)
Multiply equation (1) by 2 and add to equation 3.
$2(x + 2y + z) = 2(3) \Rightarrow 2x + 4y + 2z = 6$
$2x + 4y + 2z = 6$
$\underline{\quad x - 2y - 2z = -2}$
$\quad\quad 3x + 2y = 4 \quad$ (5)
Multiply equation (4) by 2 and add to equation (5).
$2(2x - y) = 2(5) \Rightarrow 4x - 2y = 10$
$4x - 2y = 10$
$\underline{3x + 2y = 4}$
$\quad\quad 7x = 14$
$\quad\quad\quad x = 2$
Now substitute $x = 2$ into equation (4) and solve for y.

$$2x - y = 5$$
$$2(2) - y = 5$$
$$4 - y = 5$$
$$-y = 1$$
$$y = -1$$

Finally, substitute $x = 2$ and $y = -1$ into equation (1) and solve for z.

$$x + 2y + z = 3$$
$$2 + 2(-1) + z = 3$$
$$2 - 2 + z = 3$$
$$z = 3$$

So, the solution is ordered triple $(2, -1, 3)$.

Example 3

Solve the following system using the addition method.

$$x - y + 3z = 1 \quad (1)$$
$$\underline{-x + 2y - 2z = 1} \quad (2)$$
$$x - 3y + z = 2 \quad (3)$$

Solution

Add equations (1) and (2) to eliminate x.

$$x - y + 3z = 1$$
$$\underline{-x + 2y - 2z = 1}$$
$$y + z = 2 \quad (4)$$

Add equations (2) and (3) to eliminate x.

$$-x + 2y - 2z = 1$$
$$\underline{x - 3y + z = 2}$$
$$-y - z = 3 \quad (5)$$

Now, add equations (4) and (5).

$$y + z = 2$$
$$\underline{-y - z = 3}$$
$$0 = 5 \quad \text{Which is a false statement.}$$

Therefore, the system is inconsistent and has no solution.

Note: If we obtain the true statement $0 = 0$, then the system is dependent and has an infinite number of solutions.

Exercise Set 9.4

Solve by the substitution method.

1. $x = 3$
$2x - y = 7$
$x + y + 2z = 3$

2. $3x - 2y = 7$
$y = -2$
$2x + 4z = 6$

3. $4x = 8$
$2x - y = 3$
$x + y - z = 1$

Solve the using the addition method.

4. $x + y + 3z = 0$
$2x - y + 2z = 1$
$3x + y + z = 6$

5. $3x - 2y + 4z = 22$
$x + y + z = 3$
$2x - 2y - 3z = -1$

6. $2x + 2y - z = 1$
$x + 2y - 3z = 4$
$5x + 6y - 5z = 3$

7. $x + 2y - z = -2$
$2x + 2y + 2z = 3$
$6x - 4y - 2z = 2$

8. $3x + y - z = 8$
$2x - y + 2z = 3$
$x + 2y - 3z = 5$

Answers to Exercise Set 9.4

1. $\left(3, -1, \dfrac{1}{2} \right)$

2. $(1, -2, 1)$

3. $(2, 1, 2)$

4. $(2, 1, -1)$

5. $(2, -2, 3)$

6. No solution

7. $\left(\dfrac{1}{2}, -\dfrac{1}{2}, \dfrac{3}{2} \right)$

8. Infinite number of solutions

9.5   Systems of Linear Equations: Applications and Problem Solving

Summary

Many of the applications solved in earlier chapters using only one variable can now be solved using two variables.

Example 1
Two angles are supplementary if the sum of their measures is 180°. Find two supplementary angles if one angle is 20° more than three times the other.

Solution
Let x = first angle
Let y = second angle
$x + y = 180$
$y = 3x + 20$
Solve using the substitution method.
$x + 3x + 20 = 180$
$4x + 20 = 180$
$4x = 160$
$x = 40$
$y = 3(40) + 20 = 140$
Thus, the two angles have measures of 40° and 140°.

Example 2
A boat can go upstream (against the current) a distance of 36 miles in 4 hours. It can return downstream (with the current) a distance of 45 miles in 3 hours. Determine the rate of the boat in still water and the rate of the current.

Solution
Let s = rate of boat in still water
Let c = rate of current
Use distance = rate $\cdot$ time

	Rate	Time	Distance
Upstream	$s - c$	4	36
Downstream	$s + c$	3	45

$$4(s-c) = 36 \quad \rightarrow \quad 4s - 4c = 36$$
$$3(s+c) = 45 \quad \rightarrow \quad 3x + 3y = 45$$

Solve using the addition method.
$$3(4s - 4c) = 36(3) \quad \rightarrow \quad 12s - 12c = 108$$
$$4(3s + 3c) = 45(4) \quad \rightarrow \quad \underline{12s + 12c = 180}$$
$$12s = 288$$
$$s = 12$$

Substituting $s = 12$ into the first equation,
$$4(12 - c) = 36$$
$$48 - 4c = 36$$
$$-4c = -12$$
$$c = 3$$

Thus, the rate of the boat in still water is 12 mph and the rate of the current is 3 mph.

Example 3
How many ounces of pure water and how many ounces of 16% butterfat solution should be mixed to obtain 32 ounces of a 10% butterfat solution?

Solution
Let x = number of ounces of pure water
Let y = number of ounces of 16% solution

Solution	Strength	# of ounces	Amt of butterfat
0%	0	x	$x(0)$
16%	0.16	y	$0.16y$
10%	0.10	32	$0.10(32)$

$$x + y = 32$$
$$0(x) + 0.16y = 0.10(32) \rightarrow 0.16y = 3.2 \rightarrow 16y = 320$$

Solve using the substitution method.

$$16y = 320$$
$$y = 20$$

Substituting $y = 20$ into the first equation,

$$x + 20 = 32$$
$$x = 12$$

Thus, 12 ounces on pure water and 20 ounces of 16% butterfat are needed.

This same approach can be applied to using a system with three equations and three unknowns.

Example 4

A box contains nickels, dimes, and quarters. It has 8 more dimes than nickels, and there are three less quarters than there are dimes and nickels together. The total value of the coins is $8.55. How many of each kind of coins is in the collection?

Solution

Let x = number of nickels
Let y = number of dimes
Let z = number of quarters

$$y = x + 8 \qquad\qquad (1)$$
$$z = x + y - 3 \qquad (2)$$
$$5x + 10y + 25z = 855 \qquad (3)$$

Substituting (1) and (2) into (3), we have $5x + 10(x + 8) + 25(x + y - 3) = 855$.

Now, substitute $y = x + 8$ for y into the above equation.

$$5x + 10(x + 8) + 25(x + x + 8 - 3) = 855$$
$$5x + 10x + 80 + 25(2x + 5) = 855$$
$$5x + 10x + 80 + 50x + 125 = 855$$
$$65x + 205 = 855$$
$$65x = 650$$
$$x = 10$$

Substitute $x = 10$ in equation (1) to get $y = 10 + 8 = 18$.
Substitute $x = 10$ and $y = 18$ in equation (2), $z = 10 + 18 - 3 = 25$.
Thus, the box contains 10 nickels, 18 dimes, and 25 quarters.

Exercise Set 9.5

For the following,

a. express the problem as a system of linear equations, and

b. use the method of your choice to solve the problems.

    1. Two numbers have a sum of 84. If twice the smaller number is three more than the larger number, find the numbers.

    2. How many ounces of 5% hydrochloric acid and 20% hydrochloric acid must be added together to get 10 ounces of solution that is 12.5% hydrochloric acid?

3. A person has $10,000 to invest from which he wishes to earn an annual income of $650 interest per year? He decided to bank part of the money at 5% interest and invest the remainder in bonds at $7\frac{1}{2}$% interest. How much should be invested at each rate?

4. A boat travels upstream (against the current) a distance of 18 miles in 1 hour. It returns downstream (with the current) in $\frac{3}{4}$ hour. Find the speed of the current and the speed of the boat in still water.

5. Two cars start out together and travel in opposite directions. At the end of three hours, they are 345 miles apart. If one car travels 15 mph faster than the other, what are their speeds?

6. A box has some fives, tens and twenties. It contains 30 bills worth $370. There are two more fives than tens. Find the number of each type of bill.

Answers to Exercise Set 9.5

1. a. $x + y = 84$
$2x = 3 + y$

   b. The smaller number is 29. The larger number is 55.

2. a. $x + y = 10$
$0.05x + 0.20y = 1.25$

   b. Five ounces each of the 5% and 20% solutions.

3. a. $x + y = 10,000$
$0.05x + 0.075y = 650$

   b. $4000 at 5%; $6000 at 7.5%

4. a. $\frac{3}{4}(s + c) = 18$
$1(s - c) = 18$

   b. Boat rate is 21 mph; current rate is 3 mph

5. a. $y = x + 15$
$3y + 3x = 345$

   b. One rate is 50 mph; the other rate is 65 mph

6. a. $x + y + z = 30$
$5x + 10y + 20z = 370$
$x = y + 2$

   b. 10 fives, 8 tens, 12 twenties

9.6 Solving Systems of Equations Using Matrices

Summary

A **matrix** is a rectangular array of numbers within brackets. The numbers inside the brackets are referred to as **elements** of the matrix.

Summary

For the system of equations $\quad a_1 x + b_1 y = c_1$
$$a_2 x + b_2 y = c_2$$
the **augmented** matrix is written
$$\begin{bmatrix} a_1 & b_1 & | & c_1 \\ a_2 & b_2 & | & c_2 \end{bmatrix}$$

Summary

Procedures for Row Transformations

1. Any two rows of a matrix may be interchanged. (This is the same as interchanging any two equations in the system of equations.)

2. All the numbers in a row may be multiplied (or divided) by any nonzero real number. (This is the same as multiplying both sides of an equation by a nonzero real number.)

3. All the number in a row may be multiplied by any nonzero real number. These products may then be added to the corresponding numbers in any other row. (This is equivalent to eliminating a variable from a system of equations using the addition method.)

Summary

To Write a 2 $\times$ 2 Augmented Matrix in The Form
$$\begin{bmatrix} 1 & a & | & p \\ 0 & 1 & | & q \end{bmatrix}$$

1. First, use row transformations to change the element in the first column, first row to 1.

2. Then use row transformations to change the element in the first column, second row to 0.

3. Next, use row transformations to change the element in the second column, second row to 1.

Example 2
Solve the following system of equations using matrices.
$2x + 3y = -1$
$3x + 2y = 1$

Solution
First write the augmented matrix.

$$\left[\begin{array}{cc|c} 2 & 3 & -1 \\ 3 & 2 & 1 \end{array}\right]$$

Multiply the first row by $\dfrac{1}{2}$ to obtain

$$\left[\begin{array}{cc|c} 2\left(\frac{1}{2}\right) & 3\left(\frac{1}{2}\right) & -1\left(\frac{1}{2}\right) \\ 3 & 2 & 1 \end{array}\right] \Rightarrow \left[\begin{array}{cc|c} 1 & \frac{3}{2} & -\frac{1}{2} \\ 3 & 2 & 1 \end{array}\right]$$

Next, multiply the first row by -3 and add that product to the second row. Replace the second row by the resulting sum. This results in the matrix,

$$\left[\begin{array}{cc|c} 1 & -\frac{3}{2} & -\frac{1}{2} \\ 3+1(-3) & 2+\left(-\frac{3}{2}\right)(-3) & 1+\left(-\frac{1}{2}\right)(-3) \end{array}\right] \Rightarrow \left[\begin{array}{cc|c} 1 & \frac{3}{2} & -\frac{1}{2} \\ 0 & -\frac{5}{2} & \frac{5}{2} \end{array}\right]$$

Finally, multiply the second row by $-\dfrac{2}{5}$ to obtain

$$\left[\begin{array}{cc|c} 1 & \frac{3}{2} & -\frac{1}{2} \\ 0\left(-\frac{2}{5}\right) & -\frac{5}{2}\left(-\frac{2}{5}\right) & \frac{5}{2}\left(-\frac{2}{5}\right) \end{array}\right] \Rightarrow \left[\begin{array}{cc|c} 1 & \frac{3}{2} & -\frac{1}{2} \\ 0 & 1 & -1 \end{array}\right]$$

We can now write the equivalent triangular system of equations,

$x + \dfrac{3}{2}y = -\dfrac{1}{2}$
$y = -1$

Solve by substitution. Let $y = -1$ in the top equation.

$x + \dfrac{3}{2}(-1) = -\dfrac{1}{2}$

$\qquad x - \dfrac{3}{2} = -\dfrac{1}{2}$

$\qquad\qquad x = 1$

Thus, the solution to the system is $(1, -1)$.

Summary

For a system of three linear equations we use a row transformation to write the augmented matrix in the form

$$\left[\begin{array}{ccc|c} 1 & a & b & p \\ 0 & 1 & c & q \\ 0 & 0 & 1 & r \end{array}\right]$$

Example 3
Use matrices to solve the following system of equations.

$2x - y - z = 5$
$x + 2y + 3z = -2$
$3x - 2y + z = 2$

Solution
Write the augmented matrix.

$$\begin{bmatrix} 2 & -1 & -1 & | & 5 \\ 1 & 2 & 3 & | & -2 \\ 3 & -2 & 1 & | & 2 \end{bmatrix}$$

Exchange rows 1 and 2 to obtain

$$\begin{bmatrix} 1 & 2 & 3 & | & -2 \\ 2 & -1 & -1 & | & 5 \\ 3 & -2 & 1 & | & 2 \end{bmatrix}$$

Now, multiply row 1 by –2 and add to row 2. Replace row 2 by the resulting sum.

$$\begin{bmatrix} 1 & 2 & 3 & | & -2 \\ 0 & -5 & -7 & | & 9 \\ 3 & -2 & 1 & | & 2 \end{bmatrix}$$

Next, multiply row 1 by –3 and add to row 3. Replace row 3 by the resulting sum.

$$\begin{bmatrix} 1 & 2 & 3 & | & -2 \\ 0 & -5 & -7 & | & 9 \\ 0 & -8 & -8 & | & 8 \end{bmatrix}$$

Exchange rows 2 and 3 to obtain

$$\begin{bmatrix} 1 & 2 & 3 & | & -2 \\ 0 & -8 & -8 & | & 8 \\ 0 & -5 & -7 & | & 9 \end{bmatrix}$$

Multiply row 2 by $-\dfrac{1}{8}$ to obtain

$$\begin{bmatrix} 1 & 2 & 3 & | & -2 \\ 0 & 1 & 1 & | & -1 \\ 0 & -5 & -7 & | & 9 \end{bmatrix}$$

Multiply row 2 by 5 and add to row 3. Replace row 3 by the resulting sum.

$$\begin{bmatrix} 1 & 2 & 3 & | & -2 \\ 0 & 1 & 1 & | & -1 \\ 0 & 0 & -2 & | & 4 \end{bmatrix}$$

Multiply row 3 by $-\dfrac{1}{2}$ to obtain

$$\begin{bmatrix} 1 & 2 & 3 & | & -2 \\ 0 & 1 & 1 & | & -1 \\ 0 & 0 & 1 & | & -2 \end{bmatrix}$$

Now write the equivalent triangular system of equations and solve by substitution.

$$x + 2y + 3z = -2$$
$$y + z = -1$$
$$z = -2$$

Let $z = -2$ in the second equation.

$$y + (-2) = -1$$
$$y = 1$$

Let $y = 1$ and $z = -2$ in first equation.

$$x + 2(1) + 3(-2) = -2$$
$$x - 4 = -2$$
$$x = 2$$

Finally, the solution is $(2, 1, -2)$.

Note: If row transformations yield a row of the form $\begin{bmatrix} 0 & 0 & | & K \end{bmatrix}$ or $\begin{bmatrix} 0 & 0 & 0 & | & K \end{bmatrix}$, $K \neq 0$, then the system has no solution. If row transformations yield a row of the form $\begin{bmatrix} 0 & 0 & | & 0 \end{bmatrix}$ or $\begin{bmatrix} 0 & 0 & 0 & | & 0 \end{bmatrix}$, then the system is dependent and has an infinite number of solutions.

Exercise Set 9.6

Solve the following systems using matrices.

1. $x + 3y = 7$
$-2x + y = 0$

2. $2x + 5y = 8$
$x + 2y = 3$

3. $2x + 4y = 2$
$3x + 7y = 1$

4. $2x + 5y = 4$
$3x - 6y = 33$

5. $x + 2y + 2z = 3$
$2x + 3y + 6z = 2$
$-x + y + z = 0$

6. $2x + 5y + 2z = 9$
$x + 3y + z = 4$
$2x + 3y - 3z = 1$

7. $x + 3y - 2z = 1$
$2x + 5y - 2z = 6$
$-2x - 4y + 3z = -1$

8. $5x - 9y - 9z = 7$
$3x - 4y + z = 7$
$x - 2y - z = 1$

Answers to Exercise Set 9.6

1. $(1, 2)$

2. $(-1, 2)$

3. $(5, -2)$

4. $(7, -2)$

5. $(1, 2, -1)$

6. $(5, -1, 2)$

7. $(1, 2, 3)$

8. $(5, 2, 0)$

9.7   Solving Systems of Equations Using Determinants and Cramer's Rule

Summary

Value of a Second-order Determinant

The determinant is evaluated as $\begin{vmatrix} a_1 & b_1 \\ a_2 & b_2 \end{vmatrix} = a_1 b_2 - a_2 b_1$.

Example 1

Evaluate $\begin{vmatrix} 2 & 4 \\ -3 & 7 \end{vmatrix}$

Solution

$$\begin{vmatrix} 2 & 4 \\ -3 & 7 \end{vmatrix} = 2(7) - (-3)(4) = 14 + 12 = 26$$

Summary

Cramer's Rule for Systems of Linear Equations

For a system of linear equations of the form

$$a_1 x + b_1 y = c_1$$
$$a_2 x + b_2 y = c_2$$

$$x = \frac{\begin{vmatrix} c_1 & b_1 \\ c_2 & b_2 \end{vmatrix}}{\begin{vmatrix} a_1 & b_1 \\ a_2 & b_2 \end{vmatrix}} = \frac{D_x}{D} \text{ and } y = \frac{\begin{vmatrix} a_1 & c_1 \\ a_2 & c_2 \end{vmatrix}}{\begin{vmatrix} a_1 & b_1 \\ a_2 & b_2 \end{vmatrix}} = \frac{D_y}{D}, D \neq 0$$

Example 2
Use determinants to solve the following system.
$$5x + 7y = -1$$
$$6x + 8y = 1$$

Solution
Let $a_1 = 5, \; b_1 = 7, \; c_1 = -1$
$\quad\quad a_2 = 6, \; b_2 = 8, \; c_2 = 1$

$$D = \begin{vmatrix} a_1 & b_1 \\ a_2 & b_2 \end{vmatrix} = \begin{vmatrix} 5 & 7 \\ 6 & 8 \end{vmatrix} = 5(8) - 6(7) = -2$$

$$D_x = \begin{vmatrix} c_1 & b_1 \\ c_2 & b_2 \end{vmatrix} = \begin{vmatrix} -1 & 7 \\ 1 & 8 \end{vmatrix} = (-1)(8) - (1)(7) = -15$$

$$D_y = \begin{vmatrix} a_1 & c_1 \\ a_2 & c_2 \end{vmatrix} = \begin{vmatrix} 5 & -1 \\ 6 & 1 \end{vmatrix} = (5)(1) - (6)(-1) = 11$$

$$x = \frac{D_x}{D} = \frac{15}{2} \qquad\qquad y = \frac{D_y}{D} = -\frac{11}{2}$$

Therefore, the solution to the system is the ordered pair $\left(\frac{15}{2}, -\frac{11}{2} \right)$.

Summary

Expansion of the Determinant by the Minors of the First Column

$$\begin{vmatrix} a_1 & b_1 & c_1 \\ a_2 & b_2 & c_2 \\ a_3 & b_3 & c_3 \end{vmatrix} = a_1 \begin{vmatrix} b_2 & c_2 \\ b_3 & c_3 \end{vmatrix} - a_2 \begin{vmatrix} b_1 & c_1 \\ b_3 & c_3 \end{vmatrix} + a_3 \begin{vmatrix} b_1 & c_1 \\ b_2 & c_2 \end{vmatrix}$$

Example 3

Evaluate $\begin{vmatrix} 2 & -1 & 3 \\ 4 & 0 & 7 \\ -2 & 3 & -2 \end{vmatrix}$ using expansion by the minors of the first column.

Solution

Let $\quad a_1 = 2 \qquad b_1 = -1 \quad c_1 = 3$
$\qquad a_2 = 4 \qquad b_2 = 0 \qquad c_2 = 7$
$\qquad a_3 = -2 \quad b_3 = 3 \qquad c_3 = -2$

$$\begin{vmatrix} 2 & -1 & 3 \\ 4 & 0 & 7 \\ -2 & 3 & -2 \end{vmatrix} = 2 \begin{vmatrix} 0 & 7 \\ 3 & -2 \end{vmatrix} - 4 \begin{vmatrix} -1 & 3 \\ 3 & -2 \end{vmatrix} + (-2) \begin{vmatrix} -1 & 3 \\ 0 & 7 \end{vmatrix}$$

$$= 2[(0)(-2) - (3)(7)] - 4[(-1)(-2) - (3)(3)] + (-2)[(-1)(7) - (0)(3)]$$
$$= 2[-21] - 4[-7] + (-2)[-7]$$
$$= -42 + 28 + 14$$
$$= 0$$

The determinant has a value of 0.

Summary

Cramer's Rule for a System of Equations in Three Variables

To evaluate the system
$$a_1 x + b_1 y + c_1 z = d_1$$
$$a_2 x + b_2 y + c_2 z = d_2$$
$$a_3 x + b_3 y + c_3 z = d_3$$

with
$$D = \begin{vmatrix} a_1 & b_1 & c_1 \\ a_2 & b_2 & c_2 \\ a_3 & b_3 & c_3 \end{vmatrix}, \; D_x = \begin{vmatrix} d_1 & b_1 & c_1 \\ d_2 & b_2 & c_2 \\ d_3 & b_3 & c_3 \end{vmatrix}, \; D_y = \begin{vmatrix} a_1 & d_1 & c_1 \\ a_2 & d_2 & c_2 \\ a_3 & d_3 & c_3 \end{vmatrix}, \; D_z = \begin{vmatrix} a_1 & b_1 & d_1 \\ a_2 & b_2 & d_2 \\ a_3 & b_3 & d_3 \end{vmatrix}$$

$$x = \frac{D_x}{D} \qquad y = \frac{D_y}{D} \qquad z = \frac{D_z}{D}, \; D \neq 0$$

Example 4
Solve the following system of equations using detemrinants.
$$x + y - z = -2$$
$$2x - y + z = -5$$
$$x - 2y + 3z = 4$$

Solution
Let $a_1 = 1 \quad b_1 = 1 \quad c_1 = -1 \quad d_1 = -2$
$\quad a_2 = 2 \quad b_2 = -1 \quad c_2 = 1 \quad d_2 = -5$
$\quad a_3 = 1 \quad b_3 = -2 \quad c_3 = 3 \quad d_3 = 4$

$$D = \begin{vmatrix} 1 & 1 & -1 \\ 2 & -1 & 1 \\ 1 & -2 & 3 \end{vmatrix}$$

$$= 1 \begin{vmatrix} -1 & 1 \\ -2 & 3 \end{vmatrix} - 2 \begin{vmatrix} 1 & -1 \\ -2 & 3 \end{vmatrix} - 1 \begin{vmatrix} 1 & -1 \\ -1 & 1 \end{vmatrix}$$

$$= 1(-1) - 2(1) - 1(0)$$

$$= -3$$

$$D_x = \begin{vmatrix} -2 & 1 & -1 \\ -5 & -1 & 1 \\ 4 & -2 & 3 \end{vmatrix} = -2 \begin{vmatrix} -1 & 1 \\ -2 & 3 \end{vmatrix} - (-5) \begin{vmatrix} 1 & -1 \\ -2 & 3 \end{vmatrix} - 4 \begin{vmatrix} 1 & -1 \\ -1 & 1 \end{vmatrix} = 7$$

$$D_y = \begin{vmatrix} 1 & -2 & -1 \\ 2 & -5 & 1 \\ 1 & 4 & 3 \end{vmatrix} = -22$$

$$D_z = \begin{vmatrix} 1 & 1 & -2 \\ 2 & -1 & -5 \\ 1 & -2 & 4 \end{vmatrix} = -21$$

$$x = \frac{D_x}{D} = \frac{7}{-3}, \; y = \frac{D_y}{D} = \frac{-22}{-3} = \frac{22}{3}, \; z = \frac{D_z}{D} = \frac{-21}{-3} = 7$$

The solution to the system is the ordered triple $\left(-\frac{7}{3}, \frac{22}{3}, 7\right)$.

Note: If the value of D is zero, then the system is either dependent or inconsistent. It is probably better to use another method to solve the system.

Exercise Set 9.7

Evaluate the following determinants.

1. $\begin{vmatrix} 3 & -1 \\ 2 & -6 \end{vmatrix}$

2. $\begin{vmatrix} 1 & 3 & -2 \\ -1 & -2 & -3 \\ 1 & 1 & 2 \end{vmatrix}$

Solve the system of equations using determinants.

3. $8x - 4y = 8$
    $x + 3y = 22$

4. $4x + 3y = 0$
    $3x - 4y = 25$

5. $3x + 8y = 17$
    $2x - 4y = 2$

6. $5x - 4y = 2$
    $-3x + 3y = -3$

7. $2x + 3y + 2z = 10$
    $x - y + 2z = 3$
    $x + 2y - 6z = -15$

8. $x - y + 5z = -6$
    $3x + 3y - z = 10$
    $x + 3y + 2z = 5$

9. $x + y - 3z = 1$
    $2x - y + z = 9$
    $3x + y - 4z = 8$

Answers to Exercise Set 9.7

1. -16

2. -6

3. $(4, 6)$

4. $(3, -4)$

5. $(3, 1)$

6. $(-2, -3)$

7. $(-1, 2, 3)$

8. $(1, 2, -1)$

9. $(4, 0, 1)$

Chapter 9 Practice Test

Determine, without solving the system, whether the system of equations is consistent, inconsistent, or dependent. State whether the system has exactly one solution, no solution, or an infinite number of solutions.

1. $y = \frac{3}{2}x + 2$
    $3x - 2y = 0$

2. $2x - 5y = 6$
    $3x + 2y = 3$

Solve the system of equations by the method indicated.

3. $y = 2x - 5$
    $y = x + 1$
    (Graphically)

4. $2x + y = 4$
    $3x - 2y = -1$
    (Substitution)

5. $2x + 3y = 15$
    $2x - 7y = 5$
    (Addition)

6. $5x + 4y = 12$
    $-\dfrac{3}{2}x + 2y = -2$
    (Addition)

7. $5x + 6y = -3$
    $-4x + 9y = 7$
    (Determinants)

8. $2x - 3y = 7$
    $x - 2y = 4$
    (Matrices)

9. $x - y - z = -3$
    $x - y + z = -1$
    $x + y + z = -1$
    (Addition)

10. $x - 2y + z = 2$
    $x + y + z = 8$
    $x - y - z = 2$
    (Any method)

11. a. Express the problem as a system of linear equations and

    b. use the method of your choice to find the solution to the problem.

A boat travels downstream (with the current) a distance of 40 miles in 5 hours. It takes 8 hours to make the return trip against the current. Find the rate of the boat in still water and the rate of the current.

Answers to Chapter 9 Practice Test

1. Inconsistent, no solution

2. Consistent, one solution

3. $(6, 7)$

4. $(1, 2)$

5. $(6, 1)$

6. $\left(2, \dfrac{1}{2}\right)$

7. $\left(-1, \dfrac{1}{3}\right)$

8. $(2, -1)$

9. $(-2, 0, 1)$

10. $(5, 2, 1)$

11. a. $(x + y)(5) = 40$
    $(x - y)(8) = 40$

    b. Boat rate is $= 6\dfrac{1}{2}$ mph; current rate is $1\dfrac{1}{2}$ mph

Chapter 10

10.1 Solving Linear Inequalities in One Variable

Summary

Properties Used to Solve Inequalities
1. If $a > b$, then $a + c > b + c$.
2. If $a > b$, then $a - c > b - c$.
3. If $a > b$ and $c > 0$, then $ac > bc$.
4. If $a > b$ and $c > 0$, then $\dfrac{a}{c} > \dfrac{b}{c}$.
5. If $a > b$ and $c < 0$, then $ac < bc$.
6. If $a > b$ and $c < 0$, then $\dfrac{a}{c} < \dfrac{b}{c}$.

Example 1
Solve the inequality $2x - 7 \le 1$ and graph on a number line.

Solution
$$2x - 7 \le 1$$
$$2x - 7 + 7 \le 1 + 7$$
$$2x \le 8$$
$$x \le 4$$

Summary

The solution set of an inequality graphed on the number line and written in interval notation.

Inequality	Graph	Interval Notation
$x > 5$		$(5, \infty)$
$x \geq 5$		$[5, \infty)$
$x < 5$		$(-\infty, 5)$
$x \leq 5$		$(-\infty, 5]$
$5 \leq x < 7$		$[5, 7)$
$5 \leq x \leq 7$		$[5, 7]$
$5 < x < 7$		$(5, 7)$

Example 2
Solve the inequality $2x + 8 \leq 4(x - 3)$. Give the solution on a graph and in interval notation.

Solution
$$2x + 8 \leq 4(x - 3)$$
$$2x + 8 \leq 4x - 12$$
$$-2x + 8 \leq -12$$
$$-2x \leq -20$$
$$x \geq 10$$
Graph:

Interval notation: $[10, \infty)$

Example 3
A box contains twice as many dimes as nickels and contains at least 15 coins. At least how many nickels does it have?

Solution
Let x = number of nickels
$2x$ = number of dimes
Number of nickels + Number of dimes ≥ 15
$$x + 2x \geq 15$$
$$3x \geq 15$$
$$x \geq 5$$
It must contain at least 5 nickels.

Summary

Union and Intersection of Sets

1. The **union** of set A and set B, written $A \cup B$, is the set of elements that belong to either set A or set B.

2. The **intersection** of set A and set B, written $A \cap B$, is the set of elements that are common to both set A and set B.

Example 4
Find $A \cup B$ and $A \cap B$ if $A = \{3, 4, 5, 7, 9\}$ and $B = \{5, 7, 11, 13\}$.

Solution
$A \cup B = \{3, 4, 5, 7, 9, 11, 13\}$; $A \cap B = \{5, 7\}$

Summary

A **compound inequality** is formed by joining the two inequalities with the word **and** or **or**.

To find the solution set of an inequality containing the word **and**, take the **intersection** of the solution sets of the two inequalities.

To find the solution set of an inequality containing the word **or**, take the **union** of the solution sets of the two inequalities.

Example 5
Solve $3x - 4 > 2$ and $2x - 3 \leq 5$.

Solution
Solve each inequality separately.

$$3x - 4 > 2 \qquad\qquad 2x - 3 \leq 5$$
$$3x > 6 \qquad\qquad\quad 2x \leq 8$$
$$x > 2 \qquad\qquad\quad\; x \leq 4$$

The solution is the intersection of $x > 2$ and $x \leq 4$ which is $2 < x \leq 4$.

Example 6
Solve $4x - 2 > 6$ or $2x + 5 < -3$.

Solution
Solve each inequality separately.

$$4x - 2 > 6 \qquad\qquad 2x + 5 < -3$$
$$4x > 8 \qquad\qquad\quad 2x < -8$$
$$x > 2 \qquad\qquad\quad\; x < -4$$

The solution is the union of $x > 2$ or $x < -4$ which is $x > 2$ or $x < -4$. Interval notation: $(-\infty, -4) \cup (2, \infty)$

Summary

> **Continued inequalities** are of the form $a < x < b$. In solving a continued inequality, whatever we do to one part we must do to all three parts.

Example 7
Solve the inequality $-4 \le 2(x-3) \le 2$.

Solution
$$-4 \le 2(x-3) \le 2$$
$$-4 \le 2x - 6 \le 2$$
$$2 \le 2x \le 8$$
$$1 \le x \le 4$$
Graph:

Interval notation: $[1, 4]$

Exercise Set 10.1

Express each inequality **a.** using a number line **b.** in interval notation and **c.** as a solution set.

1. $x \ge -2$

2. $x < \dfrac{2}{3}$

3. $-1 < x \le 0$

4. $x > 6$

5. $1 \le x \le \dfrac{5}{4}$

Solve the inequality and graph the solution on a number line.

6. $x + 7 > 10$

7. $5 - x > 1$

8. $6x + 3 \le 2x - 7$

9. $5(x+1) < 2(1-x)$

10. $3 \le 2x + 1 \le 5$

11. $2x > 6$ or $x - 1 < -2$

Solve the inequality and give the solution in interval notation.

12. $2x \le 1$

13. $-6x < 12$

14. $x - 7 < 5x + 4$

15. $12x + 17 \le 11 - 3x$

16. $-10 \le 5x + 5 < 5$

17. To receive a grade of C, a student must have an average greater than or equal to 70 and less than 80. If a student has scores of 65, 75, 81, 63, 60 on 5 tests, what score must be made on the 6th test to receive a grade of C?

18. Find $A \cup B$ and $A \cap B$ for sets $A = \{-2, 3, 5\}$ and $B = \{-3, -2, 4, 5, 6\}$.

Answers to Exercise Set 10.1

1. **a.** **b.** $[-2, \infty)$ **c.** $\{x|x \geq -2\}$

2. **a.** **b.** $\left(-\infty, \frac{2}{3}\right)$ **c.** $\left\{x|x < \frac{2}{3}\right\}$

3. **a.** **b.** $(-1, 0]$ **c.** $\{x|-1 < x \leq 0\}$

4. **a.** **b.** $(6, \infty)$ **c.** $\{x|x > 6\}$

5. **a.** **b.** $\left[1, \frac{5}{4}\right]$ **c.** $\left\{x|1 \leq x \leq \frac{5}{4}\right\}$

6. $x > 3$

7. $x < 4$

8. $x \leq -\frac{5}{2}$

9. $x < -\frac{3}{7}$

10. $1 \leq x \leq 2$

11. $x > 3$ or $x < -1$

12. $x \leq \frac{1}{2}; \left(-\infty, \frac{1}{2}\right]$

13. $x > -2; (-2, \infty)$

14. $x > -\frac{11}{4}; \left(-\frac{11}{4}, \infty\right)$

15. $x \leq -\frac{2}{5}; \left(-\infty, -\frac{2}{5}\right]$

16. $-3 \leq x < 0; [-3, 0)$

17. $76 \leq \text{grade} < 136$

18. $A \cup B = \{-3, -2, 3, 4, 5, 6\}$
   $A \cap B = \{-2, 5\}$

10.2 Solving Equations and Inequalities Containing Absolute Value

Summary

Procedures for Solving Equations and Inequalities Containing Absolute Value

For $a > 0$

  1. If $|x| = a$, then $x = a$ or $x = -a$.

  2. If $|x| < a$, then $-a < x < a$.

  3. If $|x| > a$, then $x < -a$ or $x > a$.

  4. If $|x| = |y|$, then $x = y$ or $x = -y$.

Example 1
Solve the equation.
$$|4x - 1| = 11$$

Solution
$$4x - 1 = 11 \quad \text{or} \quad 4x - 1 = -11$$
$$4x = 12 \quad \text{or} \quad 4x = -10$$
$$x = 3 \quad \text{or} \quad x = -\frac{5}{2}$$

Example 2
Solve the equation.
$$|2x + 3| - 2 = 9$$

Solution
First isolate the absolute value.
$$|2x + 3| - 2 = 9$$
$$|2x + 3| = 11$$
$$2x + 3 = 11 \quad \text{or} \quad 2x + 3 = -11$$
$$2x = 8 \quad \text{or} \quad 2x = -14$$
$$x = 4 \quad \text{or} \quad x = -7$$

Example 3
Solve the inequality.
$$|2x - 5| + 3 \le 12$$

Solution
Again, isolate the absolute value.

$|2x+5|+3 \le 12$

$\quad |2x+5| \le 9$

$-9 \le 2x - 5 \le 9$

$\quad -4 \le 2x \le 14$

$\quad -2 \le x \le 7$

Example 4
Solve the inequality.

$|3x+2| > 10$

Solution

$3x + 2 > 10 \quad$ or $\quad 3x + 2 < -10$

$\quad 3x > 8 \qquad\qquad 3x < -12$

$\quad x > \dfrac{8}{3} \qquad\qquad x < -4$

Example 5
Solve the equation.

$|2x+3| = |6x - 11|$

Solution

$|2x+3| = |6x - 11|$

$\quad 2x + 3 = 6x - 11 \quad$ or $\quad 2x + 3 = -(6x - 11)$

$-4x + 3 = -11 \qquad$ or $\quad 2x + 3 = -6x + 11$

$\qquad -4x = -14 \qquad$ or $\qquad 8x = 8$

$\qquad x = \dfrac{7}{2} \qquad\qquad\qquad x = 1$

Note:

If $|x| > 0$, the solution is all real numbers except $x = 0$.

If $|x| < 0$, then there is no solution.

If $|x| \ge 0$, the solution is all real numbers.

If $|x| \le 0$, the solution is $x = 0$.

Example 6
Solve $|3x+2| \le 0$.

Solution

$|3x+2| \le 0$

$\quad 3x + 2 = 0$

$\qquad x = -\dfrac{2}{3}$

Exercise Set 10.2

Solve the equation.

1. $|y+3|=1$

2. $|2x+7|=3$

3. $|4x+5|-4=4$

4. $|2s|=|s+3|$

5. $|2x+5|=|7-2x|$

Solve the inequality.

6. $|2x+5|>7$

7. $|2x-9|-1\le 0$

8. $|5x-8|>9$

9. $|5-2z|-3<0$

10. $\left|\dfrac{2x-1}{3}\right|\le 4$

11. $\left|2x-\dfrac{1}{3}\right|<2$

12. $|x+1|\ge 0$

Answers to Exercise Set 10.2

1. $-2,-4$

2. $-5,-2$

3. $-\dfrac{13}{4},\dfrac{3}{4}$

4. $-1,3$

5. $\dfrac{1}{2}$

6. $x<-6$ or $x>1$

7. $4\le x\le 5$

8. $x<-\dfrac{1}{5}$ or $x>\dfrac{17}{5}$

9. $1<z<4$

10. $-\dfrac{11}{2}\le x\le\dfrac{13}{2}$

11. $-\dfrac{5}{6}<x<\dfrac{7}{6}$

12. all real numbers

10.3 Graphing Linear Inequalities in Two Variables and Systems of Linear Inequalities

Summary

To graph a linear inequality in two variables

1. Replace the inequality symbol with an equals (=) sign.

2. Draw the graph of the linear function in step 1. If the original inequality has the symbol $\leq$ or $\geq$, the line drawn should be solid. Otherwise the line should be dashed. The graph of the line separates the coordinate plane into two half-planes, one above the line and one below the line.

3. Select **any** point not on the line (usually the origin, (0, 0) is a good choice unless it lies on the line) and determine whether its coordinates satisfy the **original inequality**. If so, shade the half-plane on the side of the line containing this point. If the point does not satisfy the inequality, shade the half-plane on the side of the line not containing this point.

Example 1
Graph $y \geq -3x + 2$.

Solution

1. Replace the inequality symbol with the equals sign: $y = -3x + 2$

2. Graph the line $y = -3x + 2$. This line has a slope of -3 and y-intercept of $(0, 2)$. The line should be solid since the original inequality contained the symbol $\geq$.

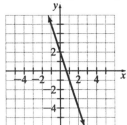

3. Choose the point $(0, 0)$ and determine if it satisfies the original inequality: $0 \geq -3(0) + 2$ is false, so the half-plane not containing the origin is shaded. Thus, the region above the line is shaded.

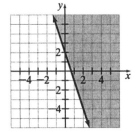

Example 2
Graph $-3x + 4y < 12$.

Solution

1. Replace the inequality symbol with an equals sign.
   $-3x + 4y = 12$

2. Graph the line $-3x + 4y = 12$ using the intercept method since the coefficients of both x and y are factors of 12. If $x = 0$, then $y = 3$, so $(0, 3)$ is the y-intercept. If $y = 0$, then $x = -4$, $(-4, 0)$ is the x-intercept. The line drawn should be dotted since original inequality contains the symbol $>$.

   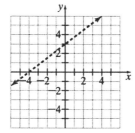

3. Select the point $(0, 0)$ (not on the line) and determine whether or not it satisfies the original inequality: $-3(0) + 4(0) < 12$ is true. Thus, shade the half-plane containing the origin. This is the half-plane below the line.

   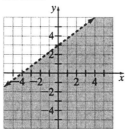

Example 3
Graph $x \geq -5$.

Solution

1. Graph the line $x = -5$. This is a vertical line 5 units to the left of the y-axis. This line should be solid since the original inequality contains the symbol $\geq$.

   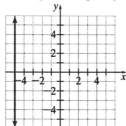

2. Since the point $(0, 0)$ does not lie on the line, substitute this point into the original inequality and determine whether it satisfies the original inequality: $0 \geq -5$ is true.

3. Shade the half-plane to the right of the line $x = -5$.

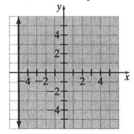

Summary

To Solve a System of Linear Inequalities

Graph each inequality on the same axes. The solution is the set of points whose coordinates satisfy all the inequalities in the system.

Example 4
Determine the solution to the system of inequalities.
$x - 3y > 6$
$2x - y \leq 0$

Solution
First graph $x - 3y > 6$.
$x - 3y > 6$
$\quad -3y > -x + 6$
$\quad\quad y < \dfrac{1}{3}x - 2$

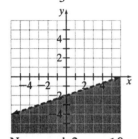

Now graph $2x - y \leq 0$ and consider the region of the plane of intersection.
$2x - y \leq 0$
$\quad -y \leq -2x$
$\quad\quad y \geq 2x$

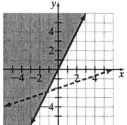

The intersection of the regions is the darkest section.

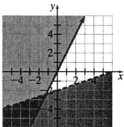

The solution is the triangular shaped region of overlap in the third quadrant.

Example 5
Determine the solution to the following system of inequalities.

$$x \geq 0$$
$$y \geq 0$$
$$x + 2y \leq 6$$
$$2x - y \geq 4$$

Solution
Since we are given the inequalities $x \geq 0$, $y \geq 0$, we restrict our attention to the first quadrant.
Graph.
$$x + 2y \leq 6$$
$$2y \leq -x + 6$$
$$y \leq -\frac{1}{2}x + 3$$

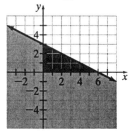

Graph.
$$2x - y \geq 4$$
$$-y \geq -2x + 4$$
$$y \leq 2x - 4$$

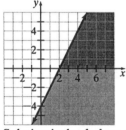

Solution is the darkest section:

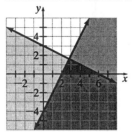

Example 6
Graph the system of inequalities.
$$|y| \le 2$$
$$|x - 1| \ge 1$$

Solution
Recall:
If $|x| < a$ and $a > 0$, then $-a < x < a$. If $|x| > a$ and $a > 0$, then $x < -a$ or $x > a$.

$|y| \le 2$ means $-2 \le y \le 2$.

$|x - 1| > 1$ means $\quad x - 1 \le -1$ or $x - 1 \ge 1$
$$\qquad\qquad\qquad\qquad x \le 0 \qquad\qquad x \ge 2$$

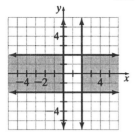

Exercise Set 10.3

Graph each inequality.

1. $x > -2$

2. $y \le -3$

3. $y < -3x + 1$

4. $y \ge \dfrac{2}{3}x + 2$

5. $-4x + 5y < 20$

6. $-3x + y > 4$

7. $\dfrac{1}{4}x - \dfrac{2}{3}y \le 2$

8. $\dfrac{1}{4}y \ge \dfrac{5}{12}x + \dfrac{3}{4}$

9. $-2x + 3y \ge 6$

10. $y \le \dfrac{3}{4}x$

11. $y > -2x$

12. $x \ge 0$

13. $y \ge 0$

14. $x > y$

Determine the solution to each system of inequalities.

15. $x - y \le 4$
   $x + y \ge 2$

16. $3x - y < 1$
   $x + 2y > 5$

17. $2x + y \ge 3$
   $x - 2y > 4$

18. $2x - 3y \le 4$
   $2x - y > -2$

19. $|x - 1| \le 3$

20. $|y + 2| > 1$

21. $|y| \ge 1$
   $|x - 1| < 2$

22. $x \ge 0$
   $y \ge 0$
   $y \le x + 1$
   $x + 2y \le 4$

23. $x \ge 0$
   $y \ge 0$
   $2x - 3y \le 6$
   $x + y \le 4$

Answers to Exercise Set 10.3

1.

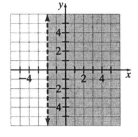

2.

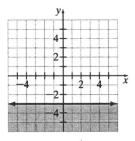

3.

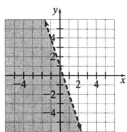

4.

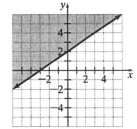

5.

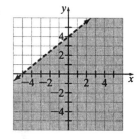

6.

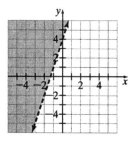

7.

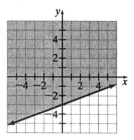

8.

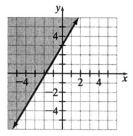

9.

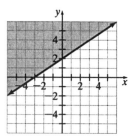

10.

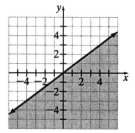

11.

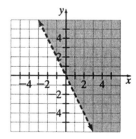

12.

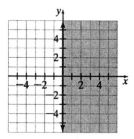

13.

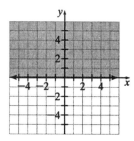

14.

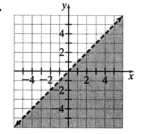

15.

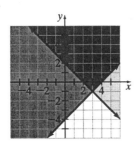

16.

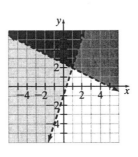

17.

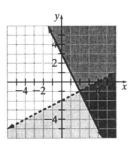

18.

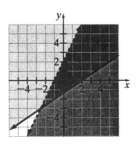

19.

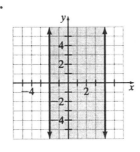

20.

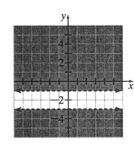

21.

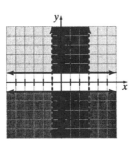

22.

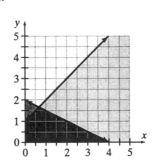

23.

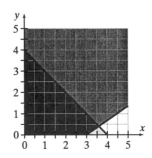

Chapter 10 Practice Test

1. Solve the inequality $\dfrac{3-4x}{2} \leq 6$ and graph the solution on the number line.

2. Solve the inequality $-2 \leq 2(x+3) \leq 6$ and write the solution in interval notation.

Find the solution to the following equations.

3. $|3x - 2| = 8$ **4.** $|4x + 2| = |x - 6|$

Find the solution to the following inequalities.

5. $\left|\dfrac{3x+1}{2}\right| \leq \dfrac{2}{3}$ **6.** $|5x + 2| - 1 > 10$

7. Graph $y \leq -3x + 2$.

8. Graph $y \geq -4$.

Graph the system of inequalities and indicate its solution.

9. $2x + y \leq 6$
    $3x - y \geq 3$

10. $|x| \leq 1$
     $|y| < 2$

Answers to Chapter 10 Practice Test

1. $x \geq -\dfrac{9}{4}$

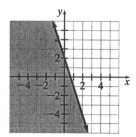

$-\dfrac{9}{4}$

2. $-4 \leq x \leq 0$; $[-4, 0]$

3. $\dfrac{10}{3}, -2$

4. $-\dfrac{8}{3}, \dfrac{4}{5}$

5. $-\dfrac{7}{9} \leq x \leq \dfrac{1}{9}$

6. $x > \dfrac{9}{5}$ or $x < -\dfrac{13}{5}$

7.

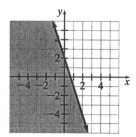

8.

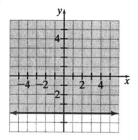

9.

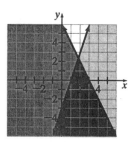

10.

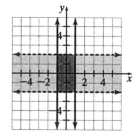

Chapter 11

11.1 Roots and Radicals

Summary

Principal Square Roots

The **principal** or **positive square root** of a positive real number a, written $\sqrt{a}$, is that positive number whose square is a. That is, $\sqrt{a} = b$ if $a = b^2$.

Example 1
Find

   a. $\sqrt{49}$ **b.** $\sqrt{\dfrac{25}{36}}$

Solution

   a. $\sqrt{49} = 7$ since $7^2 = 49$.

   b. $\sqrt{\dfrac{25}{36}} = \dfrac{5}{6}$ since $\left(\dfrac{5}{6}\right)^2 = \dfrac{25}{36}$.

Summary

_N_th Root of _a_

The nth root of a, $\sqrt[n]{a}$, where n is an **even index** and a is a positive real number, is the positive real number b such that $b^n = a$.

The nth root of a, $\sqrt[n]{a}$, where n is an **odd index** and a is any real number, is the real number b such that $b^n = a$.

Note: the nth root of a, $\sqrt[n]{a}$, is not defined if n is even and a is less than 0. For all other cases, the nth root of a is defined.

Example 2
Indicate whether or not the radical expression is a real number. If it is a real number, find its value.

a. $\sqrt[3]{27}$ **b.** $\sqrt[3]{-27}$ **c.** $\sqrt[4]{16}$

d. $\sqrt[4]{-16}$ **e.** $-\sqrt[4]{81}$

Solution

a. $\sqrt[3]{27}$ is real and equals 3.

b. $\sqrt[3]{-27}$ is real and equals –3.

c. $\sqrt[4]{16}$ is real and equals 2.

d. $\sqrt[4]{-16}$ is not a real number since the index is even but the radicand is a negative number.

e. $-\sqrt[4]{81}$ is real and equals –3.

Summary

> **For any real number a, $\sqrt{a^2} = |a|$.**

Example 3
Use absolute values to evaluate the following.

a. $\sqrt{15^2}$ **b.** $\sqrt{(-15)^2}$ **c.** $\sqrt{(-21)^2}$

d. $\sqrt{(x+7)^2}$ **e.** $\sqrt{(5-a)^2}$

Solution

a. $\sqrt{15^2} = |15| = 15$ **b.** $\sqrt{(-15)^2} = |-15| = 15$ **c.** $\sqrt{(-21)^2} = |-21| = 21$

d. $\sqrt{(x+7)^2} = |x+7|$ **e.** $\sqrt{(5-a)^2} = |5-a|$

Example 3
Simplify $\sqrt{49x^2}$ assuming x is a positive value.

Solution
$$\sqrt{49x^2} = \sqrt{(7x)^2} = 7x$$

Exercise Set 11.1

Evaluate the radical expression if it is a real number. If it is not a real number, so state.

1. $\sqrt{81}$

2. $\sqrt[3]{-125}$

3. $-\sqrt[4]{16}$

4. $\sqrt[8]{-1}$

Use absolute value to evaluate the following.

5. $\sqrt{11^2}$

6. $\sqrt{(-18)^2}$

7. $\sqrt{\left(-\dfrac{1}{2}\right)^2}$

8. $-\sqrt{(-4)^2}$

Write as an absolute value.

9. $\sqrt{(x+8)^2}$

10. $\sqrt{(10-2x)^2}$

11. $\sqrt{(x^2-1)^2}$

12. $\sqrt{(x^4+x-1)^2}$

13. Simplify $\sqrt{100y^6}$ assuming y is a positive value.

Answers to Exercise Set 11.1

1. 9

2. –5

3. –2

4. Not a real number

5. 11

6. 18

7. $\dfrac{1}{2}$

8. –4

9. $|x+8|$

10. $|10-2x|$

11. $|x^2-1|$

12. $|x^4+x-1|$

13. $10y^3$

11.2 Rational Exponents

Summary

> When a is nonnegative, n can be any index. When a is negative, n must be odd.
>
> $$\sqrt[n]{a} = a^{1/n}$$
>
> For any nonnegative number a, and integers m and n, $n \geq 2$,
>
> $$\sqrt[n]{a^m} = \left(\sqrt[n]{a}\right)^m = a^{m/n}$$

Example 1
Write the following in exponential form and simplify if possible.

   a. $\sqrt{y^3}$ **b.** $\sqrt[5]{y^{10}}$ **c.** $\sqrt[9]{x^3}$

Solution

   a. $\sqrt{y^3} = y^{3/2}$ **b.** $\sqrt[5]{y^{10}} = y^{10/5} = y^2$ **c.** $\sqrt[9]{x^3} = x^{3/9} = x^{1/3}$

Note: The rules of exponents introduced in section 1.5 are applicable with exponents that are rational numbers.

Example 2
Evaluate $\left(\dfrac{1}{27}\right)^{-1/3}$.

Solution

First, use the negative exponent rule to write $\left(\dfrac{1}{27}\right)^{-1/3} = \dfrac{1}{\left(\frac{1}{27}\right)^{1/3}}$. Now, rewrite the expression in radical form.

$$\frac{1}{\left(\frac{1}{27}\right)^{1/3}} = \frac{1}{\sqrt[3]{\frac{1}{27}}} = \frac{1}{\frac{1}{3}} = 3$$

An alternate method: The rule $\left(\dfrac{a}{b}\right)^{-n} = \left(\dfrac{b}{a}\right)^n$ can be used here.

$$\left(\frac{1}{27}\right)^{-1/3} = \left(\frac{27}{1}\right)^{1/3} = \sqrt[3]{27} = 3$$

Example 3
Use the rules of exponents to simplify the following.

a. $a^{2/3} \cdot a^{4/3}$

b. $\dfrac{c^{1/4}}{c^{-1/2}}$

c. $(x^{1/2} \cdot y^{1/4})^{-3/4}$

Solution

a. $a^{2/3} \cdot a^{4/3} = a^{\frac{2}{3}+\frac{4}{3}} = a^{6/3} = a^2$

b. $\dfrac{c^{1/4}}{x^{-1/2}} = c^{\frac{1}{4}-\left(-\frac{1}{2}\right)} = c^{\frac{1}{4}+\frac{2}{4}} = c^{3/4}$

c. $(x^{1/2} \cdot y^{1/4})^{-3/4} = x^{-3/8} \cdot y^{-3/16} = \dfrac{1}{x^{3/8} \cdot y^{3/16}}$

Note: We can factor expressions involving negative and rational exponents by factoring out the variable with the lesser exponent.

Example 4
Factor.

a. $t^{4/3} - t^{1/3}$

b. $s^{-2} + s^{-6}$

Solution

a. Since the lesser exponent is $\dfrac{1}{3}$, factor out $t^{1/3}$.

$t^{4/3} - t^{1/3} = (t^1 \cdot t^{1/3}) - t^{1/3} = t^{1/3}(t - 1)$ using the fact that $\dfrac{4}{3} = \dfrac{3}{3} + \dfrac{1}{3} = 1 + \dfrac{1}{3}$.

b. $s^{-2} + s^{-6}$
Since the lesser exponent is –6, factor out s^{-6}.
$s^{-2} - s^{-6} = (s^4 \cdot s^{-6}) - s^{-6} = s^{-6}(s^4 - 1)$ using the fact that $4 + (-6) = -2$.

Note: We can factor three-termed expressions containing rational exponents using the substitution procedure introduced in section 5.5.

Example 5

Factor $x^{2/3} - x^{1/3} - 6$.

Solution

Since $x^{2/3}$ is the square of $x^{1/3}$, substitute y for $x^{1/3}$ and write

$$x^{2/3} - x^{1/3} - 6 = (x^{1/3})^2 - x^{1/3} - 6$$
$$= y^2 - y - 6$$
$$= (y - 3)(y + 2)$$

Now, substitute $x^{1/3}$ back for each y: $(y - 3)(y + 2) = (x^{1/3} - 3)(x^{1/3} + 2)$.

Thus, $x^{2/3} - x^{1/3} - 6 = (x^{1/3} - 3)(x^{1/3} + 2)$.

Exercise Set 11.2

Write in exponential form and simplify if possible.

1. $\sqrt{2^3}$
2. $\sqrt[3]{b^5}$
3. $\sqrt{9c^4}$

4. $\sqrt[6]{x^2 y^3}$

Evaluate each of the following if possible. If the expression is not a real number, state so.

5. $9^{1/2}$
6. $27^{-2/3}$
7. $(-16)^{1/4}$

8. $16^{-5/4}$
9. $36^{1/2} + 27^{1/3}$

Simplify each of the following. Write in exponential form.

10. $y^{1/3} \cdot y^{-4/3}$
11. $\dfrac{x^{4/5}}{x^{-1/5}}$
12. $\dfrac{k^{2/3} \cdot k^{-1}}{k^{1/3}}$

13. $(8p^9 q^6)^{2/3}$
14. $\left(\dfrac{z^{10}}{x^{12}} \right)^{1/4}$
15. $\dfrac{(m^2 \cdot h)^{1/2}}{m^{3/4} \cdot h^{-1/4}}$

16. $\left(\dfrac{m^{-1/3}}{a^{-3/4}} \right)^4 (m^{-3/8} \cdot a^{1/4})^{-2}$

Factor the following.

17. $y^{-3/4} + y^{-1/4}$
18. $a^{-2} + a^{-3}$
19. $3x^{2/3} - 6x^{-4/3}$

20. $x^{2/3} + 3x^{1/3} + 2$
21. $t^{2/5} - 3t^{1/5} - 10$

Answers to Exercise Set 11.2

1. $2^{3/2}$

2. $b^{5/3}$

3. $3c^2$

4. $x^{1/3}y^{1/2}$

5. 3

6. $\dfrac{1}{9}$

7. Not a real number

8. $\dfrac{1}{32}$

9. 9

10. y^{-1} or $\dfrac{1}{y}$

11. x

12. $\dfrac{1}{k^{2/3}}$

13. $4p^6q^4$

14. $\dfrac{z^{5/2}}{x^3}$

15. $m^{1/4}h^{3/4}$

16. $\dfrac{a^{5/2}}{m^{7/12}}$

17. $y^{-3/4}(1+y^{1/2})$

18. $a^{-3}(a+1)$

19. $3x^{-4/3}(x^2-2)$

20. $(x^{1/3}+1)(x^{1/3}+2)$

21. $(t^{1/5}-5)(t^{1/5}+2)$

11.3 Multiplying and Simplifying Radicals

Summary

Product Rule for Radicals

For nonnegative real numbers a and b,

$$\sqrt[n]{a}\cdot\sqrt[n]{b}=\sqrt[n]{ab}$$

Example 1
Use the product rule to rewrite the following.

   a. $\sqrt{30}$

   b. $\sqrt[3]{y^6}$

Solution

   a. $\begin{aligned}\sqrt{30}&=\sqrt{1}\cdot\sqrt{30}\\&=\sqrt{2}\cdot\sqrt{15}\\&=\sqrt{3}\cdot\sqrt{10}\\&=\sqrt{5}\cdot\sqrt{6}\end{aligned}$

   b. $\begin{aligned}\sqrt[3]{y^6}&=\sqrt[3]{y}\cdot\sqrt[3]{y^5}\\&=\sqrt[3]{y^2}\cdot\sqrt[3]{y^4}\\&=\sqrt[3]{y^3}\cdot\sqrt[3]{y^3}\end{aligned}$

Summary

To Simplify Radicals Whose Radicands are Natural Numbers

1. Write the radicand as the product of two numbers, one of which is the largest perfect power of the given index.

2. Use the product rule to write the expression as a product of roots.

3. Find the roots of any perfect power numbers.

Note: A number is a **perfect square** if it is the square of a natural number. A number is a **perfect cube** if it is a cube of a natural number.

Example 2
Simplify the following.

   a. $\sqrt{125}$ **b.** $\sqrt[3]{81}$ **c.** $\sqrt[4]{32}$

Solution

   a. $\sqrt{125} = \sqrt{25} \cdot \sqrt{5} = 5\sqrt{5}$ **b.** $\sqrt[3]{81} = \sqrt[3]{27} \cdot \sqrt[3]{3} = 3\sqrt[3]{3}$ **c.** $\sqrt[4]{32} = \sqrt[4]{16} \cdot \sqrt[4]{2} = 2\sqrt[4]{2}$

Summary

To Simplify Radicals Whose Radicands are Variables

1. Write each variable as the product of two factors, one of which is the largest perfect power of the variable for the index.

2. Use the product rule to write the radical expression as a product of radicals. Place all perfect powers under the same radical.

3. Find the roots of any perfect powers.

Example 3
Simplify the following.

   a. $\sqrt{b^7}$ **b.** $\sqrt[3]{y^{11}}$ **c.** $\sqrt{x^5 y^6}$

Solution

a. $\sqrt{b^7} = \sqrt{b^6} \cdot \sqrt{b} = b^3 \sqrt{b}$

b. $\sqrt[3]{y^{11}} = \sqrt[3]{y^9} \cdot \sqrt[3]{y^2} = y^3 \sqrt[3]{y^2}$

c. $\sqrt{x^5 y^6} = \sqrt{x^4 y^6} \cdot \sqrt{x} = x^2 y^3 \sqrt{x}$

Summary

To Simplify Radicals

1. If the radicand contains a numerical factor, write it as a product of two numbers, one of which is the largest perfect power for the index.

2. Write each variable factor as a product of two factors, one of which is the largest perfect power of the variable for the index.

3. Use the product rule to write the radical expression as a product of radicals. Place the perfect powers (numbers and variables) under the same radical.

4. Simplify the radical containing the perfect powers.

Example 4

Simplify $\sqrt{48 c^5 d^7}$.

Solution

$$\sqrt{48 c^5 d^7} = \sqrt{16 \cdot 3 \cdot c^4 \cdot c \cdot d^6 \cdot d}$$
$$= \sqrt{16 \cdot c^4 \cdot d^6 \cdot 3cd}$$
$$= \sqrt{16 c^4 d^6} \cdot \sqrt{3cd}$$
$$= 4 c^2 d^3 \sqrt{3cd}$$

Example 5

Multiply and simplify $\sqrt[3]{4xy^5} \cdot \sqrt[3]{4x^2y}$.

Solution

First, multiply radicals and place product under a single radical. Then, simplify.

$$\sqrt[3]{4xy^5} \cdot \sqrt[3]{4x^2y} = \sqrt[3]{16x^3y^6}$$
$$= \sqrt[3]{8x^3y^6} \cdot \sqrt[3]{2}$$
$$= 2xy^2\sqrt[3]{2}$$

Exercise Set 11.3

Simplify.

1. $\sqrt{48}$

2. $\sqrt{300}$

3. $\sqrt[3]{128}$

4. $\sqrt[4]{243}$

5. $\sqrt[3]{250}$

6. $\sqrt[3]{x^5}$

7. $\sqrt{y^{11}}$

8. $\sqrt[4]{x^5y^9}$

9. $\sqrt{a^{10}b^9}$

10. $\sqrt{75y^3}$

11. $\sqrt{800k^{15}n^{10}}$

12. $\sqrt[3]{24z^5x^9}$

13. $\sqrt[4]{48x^8y^{11}}$

14. $\sqrt{8x^3y^4} \cdot \sqrt{6xy^3}$

15. $\sqrt[3]{3a^5b^6} \cdot \sqrt[3]{9a^4b^3}$

Answers to Exercise Set 11.3

1. $4\sqrt{3}$

2. $10\sqrt{3}$

3. $4\sqrt[3]{2}$

4. $3\sqrt[4]{3}$

5. $5\sqrt[3]{2}$

6. $x\sqrt[3]{x^2}$

7. $y^5\sqrt{y}$

8. $xy^2\sqrt[4]{xy}$

9. $a^5b^4\sqrt{b}$

10. $5y\sqrt{3y}$

11. $20k^7n^5\sqrt{2k}$

12. $2zx^3\sqrt[3]{3z^2}$

13. $2x^2y^2\sqrt[4]{3y^3}$

14. $4x^2y^3\sqrt{3y}$

15. $3a^3b^3$

11.4 Dividing and Simplifying Radicals

Summary

Quotient Rule for Radicals

For nonnegative real numbers a and b,

$$\frac{\sqrt[n]{a}}{\sqrt[n]{b}} = \sqrt[n]{\frac{a}{b}}, \ b \neq 0$$

Example 1
Simplify.

   a. $\dfrac{\sqrt{50}}{\sqrt{2}}$ **b.** $\dfrac{\sqrt[3]{24}}{\sqrt[3]{3}}$ **c.** $\sqrt[3]{\dfrac{125}{8}}$

   d. $\sqrt{\dfrac{8x^5}{4x^3}}$

Solution

   a. $\dfrac{\sqrt{50}}{\sqrt{2}} = \sqrt{\dfrac{50}{2}} = \sqrt{25} = 5$ **b.** $\dfrac{\sqrt[3]{24}}{\sqrt[3]{3}} = \sqrt[3]{\dfrac{24}{3}} = \sqrt[3]{8} = 2$ **c.** $\sqrt[3]{\dfrac{125}{8}} = \dfrac{\sqrt[3]{125}}{\sqrt[3]{8}} = \dfrac{5}{2}$

   d. $\sqrt{\dfrac{8x^5}{4x^3}} = \sqrt{2x^2} = \sqrt{x^2} \cdot \sqrt{2} = x\sqrt{2}$

Summary

**A Radical Expression is Simplified when the
Following are All True**

1. No perfect power are factors of the radicand.

2. No radicand contains a fraction.

3. No denominator contains a radical.

To rationalize a denominator, multiply both the
numerator and the denominator of the fraction by the
denominator, or by a radical that will result in the
radicand becoming a perfect power.

Example 2
Simplify

a. $\dfrac{2}{\sqrt{3}}$
　　　　　　　b. $\sqrt{\dfrac{1}{5}}$
　　　　　　　c. $\sqrt[3]{\dfrac{3}{4}}$

d. $\sqrt{\dfrac{6x^3y^5}{5z}}$
　　　　　e. $\sqrt[3]{\dfrac{2x}{3y^2}}$

Solution

a. $\dfrac{2}{\sqrt{3}} \cdot \dfrac{\sqrt{3}}{\sqrt{3}} = \dfrac{2\sqrt{3}}{3}$
　　b. $\sqrt{\dfrac{1}{5}} = \dfrac{\sqrt{1}}{\sqrt{5}} \cdot \dfrac{\sqrt{5}}{\sqrt{5}} = \dfrac{\sqrt{5}}{5}$
　　c. $\sqrt[3]{\dfrac{3}{4}} = \dfrac{\sqrt[3]{3}}{\sqrt[3]{4}} \cdot \dfrac{\sqrt[3]{2}}{\sqrt[3]{2}} = \dfrac{\sqrt[3]{6}}{\sqrt[3]{8}} = \dfrac{\sqrt[3]{6}}{2}$

d. $\sqrt{\dfrac{6x^3y^5}{5z}} = \dfrac{\sqrt{6x^3y^5}}{\sqrt{5z}}$

　　Now, simplify the numerator: $\sqrt{6x^3y^5} = \sqrt{x^2y^4} \cdot \sqrt{6xy} = xy^2 \cdot \sqrt{6xy}$

　　Thus, we have $\dfrac{xy^2 \cdot \sqrt{6xy}}{\sqrt{5z}}$. Now, rationalize the denominator.

　　$\dfrac{xy^2 \cdot \sqrt{6xy}}{\sqrt{5z}} \cdot \dfrac{\sqrt{5z}}{\sqrt{5z}} = \dfrac{xy^2\sqrt{30xyz}}{5z}$

e. $\sqrt[3]{\dfrac{2x}{3y^2}} = \dfrac{\sqrt[3]{2x}}{\sqrt[3]{3y^2}} \cdot \dfrac{\sqrt[3]{9y}}{\sqrt[3]{9y}} = \dfrac{\sqrt[3]{18xy}}{\sqrt[3]{27y^3}} = \dfrac{\sqrt[3]{18xy}}{3y}$

Summary

> **Rationalizing Denominators of Binomials that Contain a Radical**
>
> When the denominator of a rational expression is a binomial that contains a radical, we rationalize the denominator. We do this by multiplying both the numerator and the denominator of the fraction by the **conjugate** of the denominator. The **conjugate** of a binomial is a binomial having the same two terms with the sign of the second term changed.

Example 3
Simplify:

a. $\dfrac{4}{3+\sqrt{5}}$ **b.** $\dfrac{3}{\sqrt{7}-\sqrt{3}}$

Solution

a. $\dfrac{4}{3+\sqrt{5}}$

Multiply numerator and denominator by the conjugate of the denominator, $3-\sqrt{5}$.

$$\dfrac{4}{3+\sqrt{5}}\cdot\dfrac{3-\sqrt{5}}{3-\sqrt{5}}=\dfrac{4\left(3-\sqrt{5}\right)}{9-5}=\dfrac{4\left(3-\sqrt{5}\right)}{4}=3-\sqrt{5}$$

b. $\dfrac{3}{\sqrt{7}-\sqrt{3}}\cdot\dfrac{\sqrt{7}+\sqrt{3}}{\sqrt{7}+\sqrt{3}}=\dfrac{3\left(\sqrt{7}+\sqrt{3}\right)}{7-3}=\dfrac{3\left(\sqrt{7}+\sqrt{3}\right)}{4}$

Exercise Set 11.4

Simplify. Assume all variables represent positive real numbers.

1. $\sqrt{\dfrac{45}{5}}$ **2.** $\sqrt[3]{\dfrac{54}{2}}$ **3.** $\dfrac{\sqrt[3]{24}}{\sqrt[3]{-3}}$

4. $\dfrac{2}{3\sqrt{3}}$ **5.** $\sqrt{\dfrac{5}{6}}$ **6.** $\sqrt{\dfrac{11x}{5y}}$

7. $\dfrac{1}{\sqrt[3]{7}}$ **8.** $\dfrac{x}{\sqrt[3]{9y}}$ **9.** $\dfrac{3}{2\sqrt{x}}$

10. $\sqrt{\dfrac{8x^5y}{6z}}$ **11.** $\dfrac{\sqrt{3}}{2+\sqrt{3}}$ **12.** $\dfrac{1-\sqrt{5}}{1+\sqrt{5}}$

Answers to Exercise Set 11.4

1. 3 **2.** 3 **3.** –2

4. $\dfrac{2\sqrt{3}}{9}$ **5.** $\dfrac{\sqrt{30}}{6}$ **6.** $\dfrac{\sqrt{55xy}}{5y}$

7. $\dfrac{\sqrt[3]{49}}{7}$ **8.** $\dfrac{x\sqrt[3]{3y^2}}{3y}$ **9.** $\dfrac{3\sqrt{x}}{2x}$

10. $\dfrac{2x^2\sqrt{3xyz}}{3z}$

11. $2\sqrt{3}-3$

12. $\dfrac{\sqrt{5}-3}{2}$

11.5   Adding and Subtracting Radicals

Summary

Like radicals are radicals having the same radicand and index.

Unlike radicals are radicals differing in either the radicand or the index.

Summary

To Add or Subtract Radicals

1. Simplify each radical expression.

2. Combine like radicals (if there are any).

Example 1
Simplify.

a. $\sqrt{3}+\sqrt{48}$

b. $\sqrt[3]{16}+\sqrt[3]{54}$

c. $9\sqrt{27p^2}-4p\sqrt{108}-2\sqrt{48p^2}$

d. $\dfrac{1}{\sqrt{5}}+\sqrt{45}$

e. $\left(2\sqrt{3}-2\right)^2$

Solution

a. $\sqrt{3}+\sqrt{48}$
First, simplify each radical.
$\sqrt{3}=\sqrt{3},\ \sqrt{48}=\sqrt{16}\cdot\sqrt{3}=4\sqrt{3}$
Now, add like radicals.
$\sqrt{3}+4\sqrt{3}=5\sqrt{3}$

b. $\sqrt[3]{16}+\sqrt[3]{54}$
Simplify each radical.
$\sqrt[3]{16}=\sqrt[3]{8}\cdot\sqrt[3]{2}=2\cdot\sqrt[3]{2}$
$\sqrt[3]{54}=\sqrt[3]{27}\cdot\sqrt[3]{2}=3\cdot\sqrt[3]{2}$
Now add, $2\sqrt[3]{2}+3\sqrt[3]{2}=5\sqrt[3]{2}$.

c. $9\sqrt{27p^2} - 4p\sqrt{108} - 2\sqrt{48p^2}$

Simplify each radical.

$9\sqrt{27p^2} = 9\sqrt{9p^2} \cdot \sqrt{3} = 9 \cdot 3p\sqrt{3} = 27p\sqrt{3}$

$4p\sqrt{108} = 4p\sqrt{36} \cdot \sqrt{3} = 4p \cdot 6\sqrt{3} = 24p\sqrt{3}$

$2\sqrt{48p^2} = 2\sqrt{16p^2} \cdot \sqrt{3} = 2 \cdot 4p\sqrt{3} = 8p\sqrt{3}$

Now add, $27p\sqrt{3} - 24p\sqrt{3} - 8p\sqrt{3} = -5p\sqrt{3}$.

d. $\dfrac{1}{\sqrt{5}} + \sqrt{45}$

Simplify each radical.

$\dfrac{1}{\sqrt{5}} \cdot \dfrac{\sqrt{5}}{\sqrt{5}} = \dfrac{\sqrt{5}}{5} = \dfrac{1}{5}\sqrt{5}$

$\sqrt{45} = \sqrt{9} \cdot \sqrt{5} = 3\sqrt{5}$

Now add, $\dfrac{1}{5}\sqrt{5} + 3\sqrt{5} = \left(\dfrac{1}{5} + 3\right)\sqrt{5} = \dfrac{16}{5}\sqrt{3}$.

e. $\left(2\sqrt{3} - 2\right)^2$

First, square the binomial.

$\left(2\sqrt{3} - \sqrt{2}\right)^2 = \left(2\sqrt{3}\right)^2 - 2\left(2\sqrt{3}\right)\left(\sqrt{2}\right) + \left(-\sqrt{2}\right)^2$

Now, simplify each term.

$\left(2\sqrt{3}\right)^2 = 4 \cdot 3 = 12$

$2 \cdot 2\sqrt{3} \cdot \sqrt{2} = 4\sqrt{6}$

$\left(\sqrt{2}\right)^2 = 2$

Then add, $12 - 4\sqrt{6} + 2 = 14 - 4\sqrt{6}$.

Exercise Set 11.5

Simplify. Assume all variables represent positive real numbers.

1. $2\sqrt{48} + 3\sqrt{75}$

2. $6\sqrt{18} + \sqrt{32} - 2\sqrt{50}$

3. $2\sqrt{5} - 3\sqrt{20} - 4\sqrt{45}$

4. $2\sqrt{40} + 6\sqrt{90} - 3\sqrt{160}$

5. $3\sqrt{2x} - \sqrt{8x} - \sqrt{72x}$

6. $3\sqrt{72m^2} + 2\sqrt{32m^2} - 3m\sqrt{18}$

7. $\sqrt[3]{54} - 2\sqrt[3]{16}$

8. $2\sqrt[3]{27x} + 2\sqrt[3]{8x}$

9. $\sqrt[3]{x^2 y} - \sqrt[3]{8x^2 y}$

10. $\dfrac{4}{\sqrt{2}} + \sqrt{32}$

11. $\left(\sqrt{3} + 1\right)\left(\sqrt{5} - 1\right)$

12. $\left(2\sqrt{3} + \sqrt{5}\right)\left(3\sqrt{3} - 2\sqrt{5}\right)$

Answers to Exercise Set 11.5

1. $23\sqrt{3}$ 2. $12\sqrt{2}$ 3. $-16\sqrt{5}$

4. $10\sqrt{10}$ 5. $-5\sqrt{2x}$ 6. $17m\sqrt{2}$

7. $-\sqrt[3]{2}$ 8. $10\sqrt[3]{x}$ 9. $-\sqrt[3]{x^2 y}$

10. $6\sqrt{2}$ 11. $\sqrt{15} - \sqrt{3} + \sqrt{5} - 1$ 12. $8 - \sqrt{15}$

11.6   Solving Radical Equations

Summary

To Solve Radical Equations

1.   Rewrite the equation so that one radical containing a variable is by itself on one side of the equation.

2.   Raise each side to a power equal to the index of the radical.

3.   Collect and combine like terms.

4.   If the equation still contains a term with a variable in a radicand, repeat steps 1 through 3.

5.   Solve the resulting equation for the variable.

6.   Check all solutions in the original equation for extraneous solutions.

Example 1

Solve the equation $\sqrt{2x+1} = 3$.

Solution

$$\left(\sqrt{2x+1}\right)^2 = 3^2$$
$$2x+1 = 9$$
$$2x = 8$$
$$x = 4$$

Check: $x = 4$
$$\sqrt{2x+1} = 3$$
$$\sqrt{2(4)+1} = 3$$
$$\sqrt{9} = 3$$
$$3 = 3$$

Since $3 = 3$ is true, $x = 4$ is the solution.

Example 2

Solve the equation $\sqrt[3]{x} - 3 = 1$.

Solution

$$\sqrt[3]{x} - 3 = 1$$
$$\sqrt[3]{x} = 4$$
$$\left(\sqrt[3]{x}\right)^3 = 4^3$$
$$x = 64$$

Check: $x = 64$
$$\sqrt[3]{x} - 3 = 1$$
$$\sqrt[3]{64} - 3 = 1$$
$$4 - 3 = 1$$
$$1 = 1$$

This is a true statement so $x = 64$ is the solution.

Example 3

Solve the equation $x\sqrt{2} = \sqrt{5x-2}$.

Solution

$$\left(x\sqrt{2}\right)^2 = \left(\sqrt{5x-2}\right)^2$$
$$2x^2 = 5x - 2$$
$$2x^2 - 5x + 2 = 0$$
$$(2x-1)(x-2) = 0$$
$$2x-1 = 0 \quad \text{or} \quad x-2 = 0$$
$$2x = 1 \quad \text{or} \quad x = 2$$
$$x = \frac{1}{2} \quad \text{or} \quad x = 2$$

When these values of x are substituted into the original equation, it is found at **both** values of x make the statement true.

Thus, $x = \dfrac{1}{2}$ and $x = 2$ are solutions.

Example 4

Solve the equation $\sqrt{3x+1} = \sqrt{x+1} + 2$.

Solution

$$\left(\sqrt{3x+1}\right)^2 = \left(\sqrt{x+1} + 2\right)^2$$

$$3x+1 = x+1+4\sqrt{x+1} + 4$$

$$3x+1 = x+5+4\sqrt{x+1}$$

$$2x-4 = 4\sqrt{x+1}$$

Notice that both sides of the equation have a factor of 2: $2(x-2) = 2 \cdot 2\sqrt{x+1}$. We can divide both sides of this equation by 2 and make the equation simpler.

$$x-2 = 2\sqrt{x+1}$$

Now, square both sides to eliminate the radical.

$$(x-2)^2 = \left(2\sqrt{x+1}\right)^2$$

$$x^2 - 4x + 4 = 4(x+1)$$

$$x^2 - 4x + 4 = 4x + 4$$

$$x^2 - 8x = 0$$

$$x(x-8) = 0; \text{ or } x = 0; \ x = 8$$

Check $x = 0$.

$$\sqrt{3x+1} = \sqrt{x+1} + 2$$

$$\sqrt{3(0)+1} = \sqrt{0+1} + 2$$

$$1 = 1 + 2$$

$$1 = 3 \text{ is false.}$$

Therefore, $x = 0$ is not a solution.

Check $x = 8$.

$$\sqrt{3x+1} = \sqrt{x+1} + 2$$

$$\sqrt{3(8)+1} = \sqrt{8+1} + 2$$

$$\sqrt{25} = \sqrt{9} + 2$$

$$5 = 3 + 2 \text{ is true.}$$

Thus, $x = 8$ is the only solution.

Summary

Pythagorean Theorem

The square of the hypotenuse of a right triangle is equal to the sum of the squares of the two legs. If a and b represent the legs and c represents the hypotenuse, then

$$a^2 + b^2 = c^2$$

Example 5
A right triangle has a hypotenuse with a length of 13 units and one of the legs has a length of 5 units. Find the length of the other leg.

Solution
Use $a^2 + b^2 = c^2$ with $a = 5$ and $c = 13$.
$$5^2 + b^2 = 13^2$$
$$25 + b^2 = 169$$
$$b^2 = 144$$
$$b = \sqrt{144} = 12$$
The other leg has a length of 12 units.

Example 6
Find the area of a triangle with sides of length 3, 4, and 5 units, respectively.

Solution
Use Heron's formula: $A = \sqrt{s(s-a)(s-b)(s-c)}$, where a, b, and c are the lengths of the sides of the triangle and $s = \dfrac{a+b+c}{2}$ or $\dfrac{3+4+5}{2} = 6$.
Then, $A = \sqrt{6(6-3)(6-4)(6-5)} = \sqrt{6 \cdot 3 \cdot 2 \cdot 1} = \sqrt{36} = 6$
$A = 6$ square units

Exercise Set 11.6

Solve and check your solution(s).

1. $\sqrt{r-2} = 3$
2. $\sqrt{6k-1} = 1$
3. $\sqrt{3k+1} - 4 = 0$

4. $\sqrt{r+1} = \sqrt{2r-3}$
5. $2\sqrt{x} = \sqrt{3x+4}$
6. $k = \sqrt{k^2 + 4k - 20}$

7. $\sqrt{x^2 + 3x - 3} = x + 1$
8. $\sqrt[3]{2x+5} = \sqrt[3]{6x+1}$
9. $\sqrt[3]{x-8} + 2 = 0$

10. $\sqrt{x^2 + 12} - x = 6$
11. $\sqrt{2x+3} = 2 + \sqrt{x-2}$
12. $-\sqrt{z+5} = -\sqrt{3z+3} + 2$

13. A right triangle has legs of lengths 3 units and 6 units. What is the length of the hypotenuse?

14. A right triangle has a hypotenuse of length 8 units and a leg of length 4 units. Find the length of the other leg,

15. An equilateral triangle has a side of 6 feet. Use Heron's formula to determine its area.

Answers to Exercise Set 11.6

1. 11	**2.** $\dfrac{1}{3}$	**3.** 5
4. 4	**5.** 4	**6.** 5
7. 4	**8.** 1	**9.** 0
10. –2	**11.** 11, 3	**12.** 11
13. $3\sqrt{5}$ units	**14.** $4\sqrt{3}$ units	**15.** $9\sqrt{3}$ square units

11.7 Complex Numbers

Summary

Imaginary Numbers

$$i = \sqrt{-1},\ i^2 = -1$$

For any positive real number n, $\sqrt{-n} = i\sqrt{n}$.

Complex Numbers

Every number of the form $a + bi$ where a and b are real numbers, is a complex number.

Example 1
Write each of the following complex numbers in the form $a + bi$.

   a. $\sqrt{-16}$ **b.** $4 - \sqrt{-32}$ **c.** $3 + \sqrt{-9}$

Solution

   a. $\sqrt{-16} = i \cdot \sqrt{16} = 4i$ or $0 + 4i$

   b. $4 - \sqrt{-32} = 4 - i\sqrt{32}$
$$= 4 - i\left(4\sqrt{2}\right)$$
$$= 4 - 4\sqrt{2}i$$

c. $3 + \sqrt{-9} = 3 + i\sqrt{9}$
$$= 3 + 3i$$

Summary

To Add or Subtract Complex Numbers

1. Change all imaginary numbers to *bi* form.

2. Add (or subtract) the real parts of the complex numbers.

3. Add (or subtract) the imaginary parts of the complex numbers.

4. Write the answer in the form *a + bi*.

Example 2
Add $(5 - 6i) + (-8 + 12i)$.

Solution
$$(5 - 6i) + (-8 + 12i) = 5 - 6i - 8 + 12i$$
$$= 5 - 8 - 6i + 12i$$
$$= -3 + 6i$$

Example 3
Subtract $\left(4 + \sqrt{-27}\right) - \left(3 - \sqrt{-12}\right)$.

Solution
$$\left(4 + \sqrt{-27}\right) - \left(3 - \sqrt{-12}\right) = \left(4 + i\sqrt{27}\right) - \left(3 - i\sqrt{12}\right)$$
$$= 4 + 3\sqrt{3}i - \left(3 - 2\sqrt{3}i\right)$$
$$= 4 + 3\sqrt{3}i - 3 + 2\sqrt{3}i$$
$$= 1 + 5\sqrt{3}i$$

Summary

To Multiply Complex Numbers

1. Change all imaginary numbers to *bi* form.

2. Multiply the complex numbers as you would multiply polynomials.

3. Substitute -1 for i^2.

4. Combine the real parts and the imaginary parts. Write the answer in $a + bi$ form.

Example 4
Multiply

   a. $5i(3 + 4i)$ **b.** $(4 - 2i)(3 + 5i)$ **c.** $\left(2 - \sqrt{-5}\right)\left(3 + \sqrt{-20}\right)$

Solution

a. $\begin{aligned}5i(3 + 4i) &= 15i + 20i^2 \\ &= 15i + 20(-1) \\ &= 15i - 20 \\ &= -20 + 15i\end{aligned}$

b. $\begin{aligned}(4 - 2i)(3 + 5i) &= 12 + 20i - 6i - 10i^2 \\ &= 12 + 20i - 6i - 10(-1) \\ &= 12 + 20i - 6i + 10 \\ &= 22 + 14i\end{aligned}$

c. $\begin{aligned}\left(2 - \sqrt{-5}\right)\left(3 + \sqrt{-20}\right) &= \left(2 - \sqrt{5}i\right)\left(3 + i\sqrt{20}\right) \\ &= \left(2 - \sqrt{5}i\right)\left(3 + 2\sqrt{5}i\right) \\ &= 6 + 4\sqrt{5}i - 3\sqrt{5}i - \left(\sqrt{5}\right)\left(2\sqrt{5}\right)i^2 \\ &= 6 + 4\sqrt{5}i - 3\sqrt{5}i - \left(\sqrt{5}\right)\left(2\sqrt{5}\right)(-1) \\ &= 6 + 4\sqrt{5}i - 3\sqrt{5}i + 10 \\ &= 16 + \sqrt{5}i\end{aligned}$

Note: The **conjugate** of a complex number $a + bi$ is $a - bi$.

Summary

To Divide Complex Numbers
1. Change all imaginary numbers to *bi* form.
2. Rationalize the denominator by multiplying both the numerator and the denominator by the conjugate of the denominator.
3. Write the answer in $a + bi$ form.

Example 5
Divide.

a. $\dfrac{2 - 3i}{\sqrt{-4}}$
b. $\dfrac{3 + 2i}{4 - i}$

Solution

a.
$$\frac{2 - 3i}{\sqrt{-4}} = \frac{2 - 3i}{\sqrt{4}\,i}$$
$$= \frac{2 - 3i}{2i}$$
$$= \frac{2 - 3i}{2i} \cdot \frac{-2i}{-2i}$$
$$= \frac{-4i + 6i^2}{-4i^2}$$
$$= \frac{-6 - 4i}{4}$$
$$= \frac{2(-3 - 2i)}{2 \cdot 2}$$
$$= -\frac{3}{2} - i$$

b.
$$\frac{3 + 2i}{4 - i} \cdot \frac{4 + i}{4 + i} = \frac{12 + 3i + 8i + 2i^2}{16 - i^2}$$
$$= \frac{12 + 3i + 8i + 2(-1)}{16 - (-1)}$$
$$= \frac{10 + 11i}{17}$$
$$= \frac{10}{17} + \frac{11}{17}i$$

Exercise Set 11.7

Write each expression as a complex number in the form $a + bi$.

1. $\sqrt{-36}$

2. $4 + \sqrt{-49}$

3. $2 - \sqrt{-72}$

Perform the indicated operations.

4. $(1 - i) + (2 - 3i)$

5. $(5 + 2i) + (3 + 5i)$

6. $\left(3 + \sqrt{-9}\right) - \left(5 + \sqrt{-25}\right)$

7. $i(9i)$

8. $(1 - i)(3 + 2i)$

9. $\left(1 - \sqrt{-9}\right)\left(2 - \sqrt{-9}\right)$

10. $(3 + 4i)^2$

11. $\left(\sqrt{8} + \sqrt{-18}\right)^2$

12. $\dfrac{1}{1+i}$

13. $\dfrac{2}{1 - 3i}$

14. $\dfrac{1 + \sqrt{-4}}{1 + \sqrt{-9}}$

Answers to Exercise Set 11.7

1. $6i$ or $0 + 6i$

2. $4 + 7i$

3. $2 - 6\sqrt{2}i$

4. $3 - 4i$

5. $8 + 7i$

6. $-2 - 2i$

7. -9

8. $5 - i$

9. $-7 - 9i$

10. $-7 + 24i$

11. $-10 + 24i$

12. $\dfrac{1}{2} - \dfrac{1}{2}i$

13. $\dfrac{1}{5} + \dfrac{3}{5}i$

14. $\dfrac{7}{10} - \dfrac{1}{10}i$

Chapter 11 Practice Test

1. Use absolute value to evaluate $\sqrt{(-17)^2}$.

2. Write as an absolute value: $\sqrt{(5x + 2)^2}$.

3. Simplify $\left(\dfrac{c^{2/3} \cdot c^{-2}}{c^{-1/4}}\right)^3$.

4. Factor $3x^{2/3} + 7x^{1/3} - 6$.

Simplify. Assume all variables represent positive real numbers.

5. $\sqrt{48x^5y^6}$

6. $\sqrt[3]{9x^2y^4} \cdot \sqrt[3]{6x^3y^2}$

7. $\sqrt{\dfrac{10x^2y^5}{8z}}$

8. $\sqrt[3]{\dfrac{2}{x^2}}$

9. $\dfrac{1}{5-\sqrt{2}}$

10. $\sqrt{128} - \sqrt{32}$

11. $\sqrt[3]{-192} - 2\sqrt[3]{-375}$

12. $\left(2\sqrt{6}-1\right)\left(\sqrt{2}-\sqrt{3}\right)$

Solve the equations.

13. $\sqrt{2x-1} - 4 = 0$

14. $x = \sqrt{2x+6} - 3$

15. $\sqrt{2-7t} - 3 = \sqrt{t+3}$

16. The hypotenuse of a right triangle has a length of 12 feet. One of the legs is 6 feet. Find the length of the other leg.

17. Subtract $\left(2-\sqrt{-16}\right) - \left(3+\sqrt{-25}\right)$.

18. Multiply $\left(1+\sqrt{-4}\right)\left(5-\sqrt{-9}\right)$.

19. Divide $\dfrac{3-\sqrt{-49}}{5+\sqrt{-25}}$.

20. Evaluate $x^2 - x + 2$ if $x = -2 + i$.

Answers to Chapter 11 Practice Test

1. 17

2. $|5x+2|$

3. $\dfrac{1}{c^{13/4}}$

4. $(3x^{1/3} - 2)(x^{1/3} + 3)$

5. $4x^2y^3\sqrt{3x}$

6. $3xy^2\sqrt[3]{2x^2}$

7. $\dfrac{xy^2\sqrt{5yz}}{2z}$

8. $\dfrac{\sqrt[3]{2x}}{x}$

9. $\dfrac{5+\sqrt{2}}{23}$

10. $4\sqrt{2}$

11. $6\sqrt[3]{3}$

12. $5\sqrt{3} - 7\sqrt{2}$

13. $\dfrac{17}{2}$

14. $-3, -1$

15. -2

16. $6\sqrt{3}$ feet

17. $-1 - 9i$

18. $11 + 7i$

19. $-\dfrac{2}{5} - i$

20. $7 - 5i$

Chapter 12

12.1 Solving Quadratic Equations by Completing the Square

Summary

Square Root Property

If $x^2 = a$, where a is a real number, then $x = \pm\sqrt{a}$.

Example 1
Solve the following by using the square root property:

    a. $x^2 - 17 = 0$ **b.** $x^2 = -4$

Solution

    a. $x^2 - 17 = 0$ should be rewritten as $x^2 = 17$. Now, use the square root property and take the square root of each side of this equation: $x = \pm\sqrt{17}$. This means $x = \sqrt{17}$ and $x = -\sqrt{17}$.

    b. If $x^2 = -4$, then, using the square root property, $x = \pm\sqrt{-4}$. There are no **real solutions** to this equation. However, we can express $\sqrt{-4}$ as $2i$. Thus $2i$ and $-2i$ are our two complex number solutions.

Example 2
Solve $(x + 4)^2 - 9 = 0$.

Solution

Rewrite $(x + 4)^2 - 9 = 0$ as $(x + 4)^2 = 9$ and use the square root property: $\sqrt{(x + 4)^2} = x + 4$ and $\sqrt{9} = 3$.
Thus, we have $(x + 4) = \pm 3$; $x = -4 \pm 3$. So one value of x is $-4 + 3$ or -1. The remaining value of x is $-4 - 3$ or -7. The solutions are -1 and -7.

Summary

Pythagorean Theorem

The square of the hypotenuse of a right triangle is equal to the sum of the squares of the two legs. If a and b represent the legs and c represents and hypotenuse, then $a^2 + b^2 = c^2$.

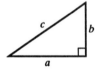

Applications

Example 3
Find the hypotenuse of a right triangle whose legs are 15 inches and 8 inches.

Solution
Let $a = 8$ and $b = 15$. Using the Pythagorean Theorem, $c^2 = a^2 + b^2$:
$$c^2 = 8^2 + 15^2$$
$$c^2 = 64 + 225$$
$$c^2 = 289$$

Using the square root property and taking the square root of both sides of this equation, we have $c = \pm\sqrt{289}$, or $c = \pm17$. Since the length of the side of a triangle cannot be negative, we drop the solution of -17 and say $c = 17$.

Summary

To Solve a Quadratic Equation by Completing the Square

1. Use the multiplication (or division) property of equality if necessary to make the leading coefficient term equal to 1.

2. Rewrite the equation with the constant isolated on the right side of the equation.

3. Take one-half the numerical coefficient of the first-degree term, square it, and add this quantity to both sides of the equation.

4. Replace the trinomial with the square of a binomial.

5. Use the square root property to take the square root of both sides of the equation.

6. Solve for the variable.

7. Check your answer in the original equation.

Example 4
Solve $x^2 - 8x + 3 = 0$ by completing the square.

Solution
Notice that the numerical coefficient of the x^2 term is already 1 so step 1 is accomplished.

2. Subtract 3 from both sides of the equation to isolate the constant on the right side.
$$x^2 - 8x = -3$$

3. Take one-half of the numerical coefficient of x term; square it and add the result to both sides of the equation.

$$\left(\frac{1}{2}\right)(-8) = -4 \text{ and } (-4)^2 = 16$$

So we have $x^2 - 8x + 16 = -3 + 16$, or $x^2 - 8x + 16 = 13$.

4. Rewrite the trinomial as the square of a binomial: $x^2 - 8x + 16 = (x - 4)^2$. Our equation now looks like $(x - 4)^2 = 13$.

5. Take the square root of both sides of this new equation: $x - 4 = \pm\sqrt{13}$.

6. Solve the equation for x: $x = 4 \pm \sqrt{13}$. The solutions are $4 + \sqrt{13}$ and $4 - \sqrt{13}$. A check shows that both values work.

Example 5

Solve $2x^2 - x = -5$ by completing the square.

Solution

Notice the numerical coefficient of the x^2 term is 2 and 2 is not equal to 1. Therefore, we must divide both sides of the equation by 2 to obtain a coefficient of 1 for x^2 term:

$$\frac{2}{2}x^2 - \frac{1x}{2} = -\frac{5}{2} \text{ or } x^2 - \frac{1}{2}x = -\frac{5}{2}$$

2. The constant term is already isolated on the right side of the equation so step 2 is not necessary.

3. Take one-half of the coefficient of the x term, square this number and add the result to both sides of the equation:

$$\frac{1}{2}\left(-\frac{1}{2}\right) = -\frac{1}{4} \text{ and } \left(-\frac{1}{4}\right)^2 = \frac{1}{16}$$

Now add $\frac{1}{16}$ to both sides of the equation:

$$x^2 - \frac{1}{2}x + \frac{1}{16} = -\frac{5}{2} + \frac{1}{16}$$

4. Replace $x^2 - \frac{1}{2}x + \frac{1}{16}$ with $\left(x - \frac{1}{4}\right)^2$. Our equation becomes $\left(x - \frac{1}{4}\right)^2 = -\frac{5}{2} + \frac{1}{16} = -\frac{40}{16} + \frac{1}{16} = -\frac{39}{16}$.

5. Take the square root of both sides of the equation:

$$\left(x - \frac{1}{4}\right) = \pm\frac{\sqrt{-39}}{\sqrt{16}}$$

$$\left(x - \frac{1}{4}\right) = \pm\frac{i\sqrt{39}}{4}$$

6. Solve for x: $x = \dfrac{1}{4} \pm \dfrac{i\sqrt{39}}{4} = \dfrac{1 \pm i\sqrt{39}}{4}$.

The solutions are $\dfrac{1}{4} + \dfrac{i\sqrt{39}}{4}$ and $\dfrac{1}{4} - \dfrac{i\sqrt{39}}{4}$.

Exercise Set 12.1

Use the square root property to solve each equation.

1. $3x^2 = 48$

2. $y^2 + 50 = 0$

3. $(x+1)^2 = 25$

4. $(x+4)^2 = -27$

5. $(x-5)^2 - 36 = 0$

6. $(3x-4)^2 = -18$

Use the Pythagorean Theorem to find the unknown lengths.

7.

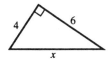

8.

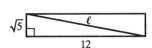

Solve each equation by completing the square.

9. $x^2 + 2x - 4 = 0$

10. $x^2 + 2x + 4 = 0$

11. $3x^2 + 4x - 6 = 0$

12. $3x^2 + 9x - 1 = 0$

Answers to Exercise Set 12.1

1. ± 4

2. $\pm 5i\sqrt{2}$

3. $-6, 4$

4. $-4 \pm 3i\sqrt{3}$

5. $-1, 11$

6. $\dfrac{4 \pm 3i\sqrt{2}}{3}$

7. $x = \sqrt{52} \approx 7.21$

8. $l = \sqrt{149} \approx 12.21$

9. $-1 \pm \sqrt{5}$

10. $-1 + i\sqrt{3}$

11. $\dfrac{-2 \pm \sqrt{22}}{3}$

12. $\dfrac{-9 \pm \sqrt{93}}{6}$

12.2   Solving Quadratic Equations by the Quadratic Formula

Summary

To Solve a Quadratic Equation by the Quadratic Formula

1. Write the quadratic equation in standard form $ax^2 + bx + c = 0$ and determine the numerical values for a, b, and c.

2. Substitute the values for a, b, c in the quadratic formula and then evaluate the formula to obtain the solution.

Quadratic Formula

$$x = \frac{-b \pm \sqrt{b^2 - 4ac}}{2a}$$

Example 1

Solve $x^2 - 6x - 16 = 0$ using the quadratic formula.

Solution

$a = 1$, $b = -6$ and $c = -16$.

$$x = \frac{-b \pm \sqrt{b^2 - 4ac}}{2a}$$

$$x = \frac{-(-6) \pm \sqrt{(-6)^2 - 4(1)(-16)}}{2(1)}$$

$$x = \frac{6 \pm \sqrt{36 + 64}}{2}$$

$$x = \frac{6 \pm \sqrt{100}}{2}$$

$$x = \frac{6 \pm 10}{2}$$

$$x = \frac{6 + 10}{2} \quad \text{or} \quad x = \frac{6 - 10}{2}$$

$$x = 8 \quad \text{or} \quad x = -2$$

The problem $x^2 - 6x - 16 = 0$ could also have been solved using factoring: $x^2 - 6x - 16 = (x - 8)(x + 2) = 0$. So $x = 8$, or $x = -2$.

A common student error is to interpret the quadratic formula as $x = -b \pm \dfrac{\sqrt{b^2 - 4ac}}{2a}$ or as $-\dfrac{b}{2a} \pm \sqrt{b^2 - 4ac}$.

Remember to divide the **entire expression** $-b \pm \sqrt{b^2 - 4ac}$ by $2a$.

Example 2
Solve $x^2 + 4 = 3x$ using the quadratic formula.

Solution
We need to write $x^2 + 4 = 3x$ in the form $ax^2 + bx + c = 0$.
$x^2 + 4 = 3x$ becomes $x^2 - 3x + 4 = 0$.
Now $a = 1$, $b = -3$, and $c = 4$. Substitute these values into the quadratic formula and evaluate it.

$$x = \frac{-b \pm \sqrt{b^2 - 4ac}}{2a}$$

$$x = \frac{-(-3) \pm \sqrt{(-3)^2 - 4(1)(4)}}{2(1)}$$

$$= \frac{3 \pm \sqrt{9 - 16}}{2}$$

$$= \frac{3 \pm \sqrt{-7}}{2}$$

$$= \frac{3 \pm i \cdot \sqrt{7}}{2}$$

The solutions are $\dfrac{3 + i\sqrt{7}}{2}$ and $\dfrac{3 - i\sqrt{7}}{2}$.

Example 3
Solve $\dfrac{1}{2}x^2 - \dfrac{3}{8}x + \dfrac{1}{4} = 0$.

Solution
When a quadratic equation contains fractions, it is advisable to multiply both sides of the equation by the LCD of all fractions that appear in the equation. This will clear all fractions in the equation.

$$8\left(\frac{1}{2}x^2 - \frac{3}{8}x + \frac{1}{4}\right) = 8(0)$$

$$4x^2 - 3x + 2 = 0$$

Now $a = 4$, $b = -3$, and $c = 2$.

$$x = \frac{-b \pm \sqrt{b^2 - 4ac}}{2a}$$

$$x = \frac{-(-3) \pm \sqrt{(-3)^2 - 4(4)(2)}}{2(4)}$$

$$= \frac{3 \pm \sqrt{9 - 32}}{8}$$

$$= \frac{3 \pm \sqrt{-23}}{8}$$

$$= \frac{3 \pm i \cdot \sqrt{23}}{8}$$

The solutions are $\dfrac{3+i\sqrt{23}}{8}$ and $\dfrac{3-i\sqrt{23}}{8}$.

If we are given the solutions of a quadratic equation, then we can find the factors of the quadratic equation which gives us the quadratic equation, upon multiplication.

Example 4
Write an equation whose solutions are -4 and 5.

Solution
If -4 and 5 are the solutions of a quadratic equation, then $(x + 4)(x - 5)$ are the factors. Multiplying out $(x + 4)(x - 5)$ gives $x^2 - 5x + 4x - 20$ or $x^2 - x - 20$.
The equation is $x^2 - x - 20 = 0$.

Summary

Solutions of a Quadratic Equation

For a quadratic equation of the form $ax^2 + bx + c = 0$ with a not equal to zero:

If $b^2 - 4ac > 0$, then the quadratic equation has two distinct real number solutions.

If $b^2 - 4ac = 0$, then the quadratic equation has a single real number solution.

If $b^2 - 4ac < 0$, then the quadratic equation has no real number solution.

Note: $b^2 - 4ac$ is called the discriminant.

Example 5
Use the discriminant to determine the nature of the solutions of the equation $2x^2 + 5 = 4x$.

Solution
This equation is equivalent to $2x^2 - 4x + 5 = 0$, in which $a = 2$, $b = -4$, and $c = 5$. The value of the discriminant is given by $b^2 - 4ac = (-4)^2 - 4(2)(5) = -24$. Since this value is less than zero, the quadratic equation has no real number solution.

Example 6
Let $f(x) = 2x^2 + x + 6$. Find the values of x for which $f(x) = 8$.

Solution

Since $f(x) = 2x^2 + x + 6$ and $f(x) = 8$, the equation becomes $8 = 2x^2 + x + 6$, or $0 = 2x^2 + x - 2$.
Use the quadratic formula with $a = 2$, $b = 1$, and $c = -2$ to get

$$x = \frac{-b \pm \sqrt{b^2 - 4ac}}{2a}$$

$$= \frac{-1 \pm \sqrt{1^2 - 4(2)(-2)}}{2(2)}$$

$$= \frac{-1 \pm \sqrt{17}}{4}$$

The values of x are $\dfrac{-1 + 17}{4}$ and $\dfrac{-1 - \sqrt{17}}{4}$.

Example 7

An apple thrown from a height of 10 feet straight up into the air with an initial velocity of 78 feet per second reaches a distance (in feet) above the ground after t seconds given by the formula $h = 10 + 78t - 16t^2$.

   a. When does the apple reach a height of 100 feet?

   b. When does the apple hit the ground?

Solution

   a. When the apple reaches 100 feet, the h value is 100. We have

$$100 = 10 + 78t - 16t^2$$
$$0 = -90 + 78t - 16t^2$$
$$0 = -16t^2 + 78t - 90$$

Now, $a = -16$, $b = 78$, and $c = -90$.
Use the quadratic formula:

$$t = \frac{-b \pm \sqrt{b^2 - 4ac}}{2a}$$

$$= \frac{-78 \pm \sqrt{78^2 - 4(-16)(-90)}}{2(-16)}$$

$$= \frac{-78 \pm \sqrt{6084 - 5760}}{-32}$$

$$= \frac{-78 \pm \sqrt{324}}{-32}$$

$$= \frac{-78 \pm 18}{-32}$$

Then, $t = \dfrac{-78 + 18}{-32}$ or $t = \dfrac{-78 - 18}{-32}$

$\qquad t = \dfrac{-60}{-32}$ or $t = \dfrac{-96}{-32}$

$\qquad t = 1.875$ or $t = 3$

Thus, the apple reaches a height of 100 feet after 1.875 seconds and again after 3 seconds.

b. When the apple hits the ground, the h value is zero. The equation becomes $0 = 10 + 78t - 16t^2$ or $0 = -16t^2 + 78t + 10$.

Now, $a = -16$, $b = 78$, and $c = 10$.

Use the quadratic formula.

$$t = \frac{-b \pm \sqrt{b^2 - 4ac}}{2a}$$

$$= \frac{-78 \pm \sqrt{78^2 - 4(-16)(10)}}{2(-16)}$$

$$= \frac{-78 \pm \sqrt{6084 + 640}}{-32}$$

$$= \frac{-78 \pm \sqrt{6724}}{-32}$$

$$= \frac{-78 \pm 82}{-32}$$

Then, $t = \dfrac{-78 + 82}{-32}$ or $t = \dfrac{-78 - 82}{-32}$

$\quad\quad t = \dfrac{4}{-32}$ or $t = \dfrac{-160}{-32}$

$\quad\quad t = -0.125$ or $t = 5$

Since t cannot be a negative number, the reasonable value is $t = 5$ seconds.

Exercise Set 12.2

Solve the given equations using the quadratic formula.

1. $y^2 + 3y - 4 = 0$ **2.** $x^2 + 3x - 2 = 0$ **3.** $5t^2 + 33t = 14$

4. $2x^2 - 3x + 2 = 0$ **5.** $x^2 - \dfrac{7}{6}x + \dfrac{2}{3} = 0$

Write an equation whose solutions are

6. $\dfrac{1}{2}, 3$ **7.** $\dfrac{3}{5}, \dfrac{1}{4}$

8. Set $g(x) = x^2 - 5x + 9$. Find the values of x for which $g(x) = 2$.

9. Set $p(a) = 5x^2 + 2x$. Find the values of a for which $p(a) = 1$.

10. An object is thrown upward with an initial velocity of 60 feet per second. The distance above the ground, h, after t seconds is given by $h = 60t - 16t^2$. When does the object reach a height of 56 feet.

Answers to Exercise Set 12.2

1. $-4, 1$

2. $\dfrac{-3 \pm \sqrt{17}}{2}$

3. $-7, \dfrac{2}{5}$

4. $\dfrac{3 \pm i\sqrt{7}}{4}$

5. $\dfrac{7 \pm i\sqrt{47}}{12}$

6. $2x^2 - 7x + 3$

7. $20x^2 - 17x + 3$

8. $\dfrac{5 \pm i\sqrt{3}}{2}$

9. $\dfrac{-1 \pm \sqrt{6}}{5}$

10. 1.75 seconds, 2 seconds

12.3 Quadratic Equations: Applications and Problem Solving

One goal of this section is to apply quadratic equations to real life situations.

Motion Problems

Example 1
Bob drove 20 miles at a constant speed then increased his speed by 10 miles per hour for the next 30 miles. If the time required to travel the 50 miles was 0.9 hours, find the speed he drove during the first 20 minutes.

Solution
Use $d = r \cdot t$ and orgainize the information in a chart: Let x represent the speed during the first 20 miles of the trip. Recall, $t = \dfrac{d}{r}$.

d	r	t
20 miles	x (mph)	$\dfrac{20}{x}$
30 miles	$x + 10$	$\dfrac{30}{x+10}$

Since the total time of travel is 0.9 hours, the equation is $\dfrac{20}{x} + \dfrac{30}{x+10} = 0.9$. The LCD is $x(x + 10)$ so the equation can be rewritten as

$$\frac{20(x+10)}{x(x+10)} + \frac{30x}{(x+10)x} = 0.9$$

$$\frac{20x + 200 + 30x}{x(x+10)} = \frac{9}{10}$$

$$\frac{50x + 200}{x(x+10)} = \frac{9}{10}$$

Cross multiply: $10(50x + 200) = 9x(x + 10)$

$$500x + 2000 = 9x^2 + 90x$$

$$0 = 9x^2 - 410x - 2000$$

Use quadratic formula: $x = \dfrac{-(-410) \pm \sqrt{(-410)^2 - 4(9)(-2000)}}{2(9)}$

$$x = 50 \text{ miles per hour or } x = -4.44 \text{ miles per hour}$$

Since x cannot be negative, the only answer which makes sense is $x = 50$ miles per hour.

Work Problems

Example 2
Two painters take 6 hours to paint a room when they work together. If they worked alone, the more experienced painter could complete the job 1 hour faster than the less experienced painter. How long would it take each of them to paint the room working alone?

Solution
Let x = number of hours needed by the less experienced painter to paint the room. Then $x - 1$ is the number of hours needed by the more experienced painter.

In **1 hour**, the less experienced painter can paint $\dfrac{1}{x}$ of the room. Also in 1 hour, the more experienced painter

can paint $\dfrac{1}{x-1}$ of the room. Finally, in 1 hour working together, **both painters** can paint $\dfrac{1}{6}$ of the room. Then

the portion of the job each painter can do in an hour **combined** should equal $\dfrac{1}{6}$ or $\dfrac{1}{x} + \dfrac{1}{x-1} = \dfrac{1}{6}$. Multiply both

sides by the LCD or $6x(x - 1)$:

$$6x(x-1)\left(\frac{1}{x} + \frac{1}{x-1}\right) = \frac{1}{6} \cdot 6x(x-1)$$

$$6(x-1) + 6x = x(x-1)$$

$$6x - 6 + 6x = x^2 - x$$

$$12x - 6 = x^2 - x$$

$$0 = x^2 - 13x + 6$$

Using the quadratic formula,

$$x = \frac{-(-13) \pm \sqrt{(-13)^2 - 4(1)(6)}}{2(1)} = \frac{13 \pm \sqrt{145}}{2}$$

$x = 12.52$
or $x \approx 0.48$.
The solution of 0.48 does not make sense, so the only solution is 12.52 hours.
It takes the painters 12.52 hours and 13.52 hours to paint the room working alone.

Summary

When the square of a variable appears in a formula, you may need to use the square root property to solve for the variable. However, when you use the square root property in most formulas, you will use only the positive or principle root.

Example 3

The surface area of a sphere is $A = \pi r^2$ where A is the area and r is the radius.

    a. Find the the surface area when r is 5 cm.

    b. Solve the equation for r.

Solution

    a. Substitute $r = 5$ to obtain

$$A = \pi(5)^2$$
$$= \pi(25)$$
$$= 25\pi \approx 78.54 \text{ sq. cm}$$

    b. $A = \pi r^2$

$$\frac{A}{\pi} = r^2 \quad \text{Divide by } \pi.$$

$$\sqrt{\frac{A}{\pi}} = r \quad \text{Take square root.}$$

Example 4

Solve $p = \sqrt{a^2 - b^2}$ for a.

Solution

$$p = \sqrt{a^2 - b^2}$$
$$p^2 = a^2 - b^2 \quad \text{Square both sides.}$$
$$p^2 + b^2 = a^2 \quad\quad \text{Add } b^2 \text{ to both sides.}$$
$$\sqrt{p^2 + b^2} = a \quad\quad \text{Take square root.}$$

Exercise Set 12.3

    1. Two molding machines can complete an order in 12 hours. The larger machine can complete the order by itself in one hour less time than the smaller machine can by itself. How long will it take each machine to complete the order working by itself.

2. Carmen canoed downstream going with the current for 3 miles, then turned around and canoed upstream against the current back to the starting point. If the total time she spent canoeing was 4 hours, and the speed of the current was 0.4 miles per hour, what is the speed of the canoe in still water.

3. Solve $F = \dfrac{1}{3}mv^2$ for v.

4. Solve $a^2 + b^2 = c^2$ for a.

5. Solve $p = \sqrt{b^2 - 7c + d^2}$ for d.

Answers to Exercise Set 12.3

1. Larger machine in 23.51 hours, smaller machine in 24.51 hours

2. 1.6 miles per hour

3. $v = \sqrt{\dfrac{3F}{m}}$

4. $a = \sqrt{c^2 - b^2}$

5. $d = \sqrt{p^2 - b^2 + 7c}$

12.4 Factoring Expressions and Solving Equations that are Quadratic in Form

Summary

To Solve Equations Quadratic in Form

1. If necessary rewrite the equation in descending order of the variable with one side of the equation equal to zero.

2. Rewrite the variable in the highest degree term as the square of the variable in the middle term.

3. Make a substitution that will result in an equation of the form $au^2 + bu + c = 0$, where u is a function of the original variable.

4. Solve the equation $au^2 + bu + c = 0$ by factoring, by the quadratic formula, or by completing the square.

5. Substitute the original variable expression for u and solve the resulting equation for the original variable.

6. Check your solutions.

Example 1

Solve $x^4 + 3x^2 = 18$.

Solution

1. $x^4 + 3x^2 - 18 = 0$ (standard form)

2. $(x^2)^2 + 3(x^2) - 18 = 0$

3. Let $u = x^2$. Then the equation becomes $u^2 + 3u - 18 = 0$.

4. Solve the equation by factoring: $(u + 6)(u - 3) = 0$
   Thus, $u + 6 = 0$ or $u - 3 = 0$
              $u = -6$ or $u = 3$

5. $u = -6$ becomes $x^2 = -6$ so that $x = \pm\sqrt{-6} = \pm i\sqrt{6}$
   $u = 3$ becomes $x^2 = 3$ so that $x = \pm\sqrt{3}$

6. A check shows all four values work. The solutions are $i\sqrt{6}$, $-i\sqrt{6}$, $\sqrt{3}$, and $-\sqrt{3}$.

Example 2

Solve $x^{1/2} - 2x^{1/4} = 8$.

Solution

1. Rewrite the equation with the right side equal to zero: $x^{1/2} - 2x^{1/4} - 8 = 0$.

2. $(x^{1/4})^2 - 2x^{1/4} - 8 = 0$

3. Let $u = x^{1/4}$ so that the equation becomes $u^2 - 2u - 8 = 0$.

4. Solve by factoring,
$$u^2 - 2u - 8 = 0$$
$$(u - 4)(u + 2) = 0$$
$$u - 4 = 0 \quad \text{or} \quad u + 2 = 0$$
$$u = 4 \quad \text{or} \qquad u = -2$$

5. $u = 4$ becomes $x^{1/4} = 4$ so that $x = 4^4 = 256$ upon raising to the fourth power.
   $u = -2$ becomes $x^{1/4} = -2$ so that $x = (-2)^4 = 16$.

6. A check shows both values work.
   The solutions are 16 and 256.

Example 3

Solve $2 + \dfrac{5}{x^2} = \dfrac{6}{x}$.

Solution

Multiply both sides by x^2 to clear fractions.

$$x^2\left(2+\frac{5}{x^2}\right)=x^2\left(\frac{6}{x}\right)$$

$$2x^2+5=6x$$

$$2x^2-6x+5=0$$

To solve, use the quadratic formula.

$$x=\frac{-b\pm\sqrt{b^2-4ac}}{2a}$$

$$=\frac{-(-6)\pm\sqrt{(-6)^2-4(2)(5)}}{2(2)}$$

$$=\frac{6\pm\sqrt{36-40}}{4}$$

$$=\frac{6\pm\sqrt{-4}}{4}$$

$$=\frac{6\pm2i}{4}$$

$$=\frac{3\pm i}{2}$$

Exercise Set 12.4

Solve the given equations.

1. $3=\dfrac{-4}{x}+\dfrac{15}{x^2}$

2. $2x-\dfrac{20}{x}=0$

3. $2=\dfrac{3}{x}+\dfrac{1}{x^2}$

4. $y^4-4y^2+3=0$

5. $x^4-3x^2-4=0$

6. $x^{2/3}-2x^{1/3}-3=0$

7. $x^{1/2}-4x^{1/4}+3=0$

8. $x-13\sqrt{x}+40=0$

Answers to Exercise Set 12.4

1. $-3,\ \dfrac{5}{3}$

2. $\pm\sqrt{10}$

3. $\dfrac{3\pm\sqrt{17}}{4}$

4. $\pm1,\ \pm\sqrt{3}$

5. $\pm2,\ \pm i$

6. $-1,27$

7. $1,81$

8. $25,64$

12.5 Graphing Quadratic Functions

Summary

Graphing Quadratic Functions

The graph of $f(x) = ax^2 + bx + c$ *(or* $y = ax^2 + bx + c$*)*
for $a \neq 0$ is a **parabola**. The parabola opens upward if $a > 0$ and opens downward if $a < 0$.

The **axis of symmetry** is $x = -\dfrac{b}{2a}$.

The **vertex** is $\left(-\dfrac{b}{2a},\, f\left(-\dfrac{b}{2a}\right)\right)$ or $\left(-\dfrac{b}{2a},\, \dfrac{4ac - b^2}{4a}\right)$.

To find the **y-intercept**, set $x = 0$ and solve for y.

To find the **x-intercept**, set $f(x) = 0$ (or $y = 0$) and solve for x.

Example 1
Consider the function $f(x) = x^2 + 2x - 3$.

    a. Determine whether this parabola opens upward or downward.

    b. Find the y-intercept.

    c. Find the vertex.

    d. Find the x-intercepts (if they exist).

    e. Sketch the graph.

Solution
Since $f(x) = x^2 + 2x - 3$, then $a = 1$, $b = 2$, and $c = -3$.

    a. Since $a > 0$, the parabola opens upward.

    b. To find the y-intercept, substitue zero for x and solve the resulting equation for y: $y = 0^2 + 2(0) - 3 = -3$.
       Thus, $(0, -3)$ is the y-intercept.

    c. To find the x coordinate of the vertex, the formula $x = -\dfrac{b}{2a}$ can be used: $x = -\dfrac{2}{2(1)} = -1$. Substitute
       $x = -1$ in the equation and solve for y: $y = (-1)^2 + 2(-1) - 3 = -4$. Thus, the vertex is located at $(-1, -4)$.

    d. To find the *x*-intercepts, substiutte *y* = 0 in the equation and solve the resulting equation for *x*:

$$0 = x^2 + 2x - 3$$
$$0 = (x+3)(x-1)$$

Thus, $x = -3$ or $x = 1$.

    e. The graph can now be sketched using the information from parts a–d.

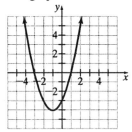

Example 2

Consider the function $g(x) = -x^2 + 2x + 2$.

    a. Determine whether this parabola opens upward or downward.

    b. Find the axis of symmetry.

    c. Find the vertex.

    d. Find the *y*-intercept.

    e. Sketch a graph.

Solution

Since $g(x) = -x^2 + 2x + 2$. Then $a = -1$, $b = 2$, and $c = 2$.

    a. Since $a < 0$, the parabola opens downward.

    b. Axis of symmetry: Use $x = -\dfrac{b}{2a} = -\dfrac{2}{2(-1)} = 1$. The parabola will be symmetric about the line $x = 1$.

    c. The *x* coordinate of the vertex is 1. Evaluate *f*(1) to find the *y* coordinate of the vertex:

        $f(1) = -(1)^2 + 2(1) + 2 = 3$; vertex is (1, 3).

    d. To find the *y*-intercept, let *x* = 0 and solve.

        $y = -(0)^2 + 2(0) + 2 = 2$

e. The graph can be sketched using the information from parts a–d.

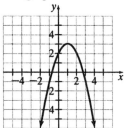

Summary

Maximum and Minimum Values

A parabola that opens upward has a **minimum value** at its vertex.

A parabola that opens downward has a **maximum value** at its vertex.

Given $f(x) = ax^2 + bx + c,$ the maximum or minimum value will occur at $-\dfrac{b}{2a}$ and the value will be

$\dfrac{4ac - b^2}{4a}$. There are many real-life applications that require finding minimum and maximum values.

Example 3
Find the maximum value of $f(x) = 2x^2 - 4x + 17$.

Solution
Let $a = 2$, $b = -4$, and $c = 17$.
The maximum value occurs at $x = -\dfrac{b}{2a} = -\dfrac{-4}{2(2)} = 1$ and the value is

$\dfrac{4ac - b^2}{4a} = \dfrac{4(2)(17) - (-4)^2}{4(2)} = \dfrac{136 - 16}{8} = \dfrac{120}{8} = 15.$

Also, note that $f(1) = 2(1)^2 - 4(1) + 17 = 2 - 4 + 17 = 15$ gives the maximum value.

Example 4
An object is thrown upward with an initial velocity of 192 feet per second. The distance of the object above the ground, d, after t seconds is given by the formula $d(t) = -16t^2 + 192t$. Find the maximum height of the object.

Solution
Let $a = -16$, $b = 192$, and $c = 0$.

The maximum value occurs at $t = -\dfrac{b}{2a} = -\dfrac{192}{2(-16)} = \dfrac{192}{32} = 6$ seconds.

The maximum value is $\dfrac{4ac - b^2}{4a} = \dfrac{4(-16)(0) - (192)^2}{4(-16)} = \dfrac{-36,864}{-64} = 576$ feet.

The maximum height is 576 feet.

Summary

Translations of Parabolas

For any function $f(x) = ax^2$, the graph of $g(x) = a(x - h)^2 + k$ will have the same shape as the graph of $f(x)$. The graph of $g(x)$ will be the graph of $f(x)$ shifted as follows:

- If h is a positive real number, the graph will be shifted h units to the right.

- If h is a negative real number, the graph will be shifted $|h|$ units to the left.

- If k is a positive real number, the graph will be shifted k units up.

- If k is a negative real number, the graph will be shifted $|k|$ units down.

The graph of any function of the form $f(x) = a(x - h)^2 + k$ will be a parabola with axis of symmetry $x = h$ and vertex at (h, k).

Example 5
Graph.

     a. $f(x) = (x - 1)^2 + 5$ **b.** $g(x) = -(x + 1)^2 + 2$

Solution

a. The graph of $f(x) = (x-1)^2 + 5$ has the same shape as $y = x^2$ shifted 1 unit to the right and shifted up 5 units.

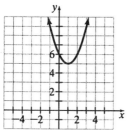

b. The graph of $g(x) = -(x+1)^2 + 2$ has the same shape as $y = -x^2$ shifted 1 unit to the left and shifted up 2 units.

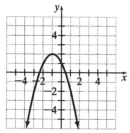

Example 6

Graph $f(x) = x^2 + 6x + 7$ by expressing $f(x)$ in the form $f(x) = a(x-h)^2 + k$.

Solution
Use the method of completing the square.
$$f(x) = x^2 + 6x + 9 - 9 + 7$$
$$= (x^2 + 6x + 9) - 2$$
$$= (x+3)^2 - 2$$

The graph has the same shape as $y = x^2$ shifted 3 units to the left and shifted down 2 units.

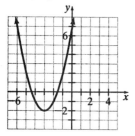

Exercise Set 12.5

For exercises 1–2,

    a. Determine whether the parabola opens upward or downward.

    b. Find the *y*-intercept.

    c. Find the vertex.

    d. Find the *x*-intercepts (if they exist).

    e. Sketch the graph.

    1. $f(x) = 2x^2 - 6x + 4$ **2.** $y = -2x^2 + 5x + 4$

Find the maximum or minimum value.

    3. $f(x) = 2x^2 - 8x + 3$ **4.** $g(x) = -4x^2 + 8x + 10$

    5. Graph $p(x) = x^2 - 4x + 5$.

Answers to Exercise Set 12.5

    1. a.   Upward   **b.** 4   **c.** $\left(\dfrac{3}{2}, -\dfrac{1}{2} \right)$

         d. $x = 1, x = 2$ **e.**

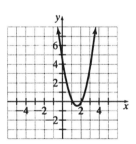

    2. a.   Downward   **b.** 4   **c.** $\left(\dfrac{5}{4}, \dfrac{57}{8} \right)$

d. $\left(\dfrac{5 \pm \sqrt{57}}{4} \right)$ **e.**

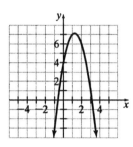

3. –5   **4.** 14   **5.**

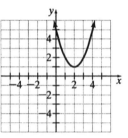

12.6 Quadratic and Other Inequalities in One Variable

Summary

Solutions of a Quadratic Inequality
1. The **solution** to a quadratic inequality is the set of all values that make the inequality a true statement.
2. One method used to solve a quadratic inequality is a **sign graph**.

Example 1
Solve $x^2 - 3x - 10 > 0$.

Solution

1. Factor $x^2 - 3x - 10$. $x^2 - 3x - 10 > 0$ is equivalent to $(x - 5)(x + 2) > 0$.

2. Find the **boundary values**. These are the values that make each factor equal to 0. If $x - 5 = 0$, then $x = 5$. If $x + 2 = 0$, then $x = -2$.

3. Draw two number lines together; one for each factor and label the boundary values. Draw vertical lines through these values.

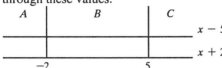

4. Notice that the two vertical lines separate the number line into three regions which are labeled A, B and C. The sign of each factor is determined for each region as well as the sign of the product of the factors. (The sign of the product is in parentheses).

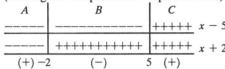

5. Since we are solving $x^2 - 3x - 10 > 0$, the solutions are the numbers in the regions A and C. In interval notation, $(-\infty, -2) \cup (5, \infty)$. The boundary values are **not included** because the inequality is >.

The solution in set builder notation is $\{x | x < -2 \text{ or } x > 5\}$.

The solution can be graphed on the number line as follows:

An alternate method of solving a quadratic inequality uses a single number line. For this, set up a number line with the boundary values.

$$\begin{array}{ccc} A & B & C \\ \hline \quad -2 \quad & \quad 5 \quad & \end{array}$$

The boundaries form three regions (intervals). Select a test value on each interval and see if it satisfies the original inequality.

Interval A: Use $x = -4$.   Interval B: Use $x = 1$.   Interval C: Use $x = 7$.

Is $x^2 - 3x - 10 > 0$?   Is $x^2 - 3x - 10 > 0$?   Is $x^2 - 3x - 10 > 0$?

$(-4)^2 - 3(-4) - 10 > 0$ $1^2 - 3(1) - 10 > 0$ $7^2 - 3(7) - 10 > 0$

$\qquad\qquad 18 > 0$ $\qquad\qquad -12 > 0$ $\qquad\qquad 18 > 0$

$\qquad\qquad$ true $\qquad\qquad$ false $\qquad\qquad$ true

The solution is $\{x | x < -2 \text{ or } x > 5\}$. In interval notation, it is $(-\infty, -2) \cup (5, \infty)$.

Summary

To Solve Quadratic and Other Inequalities

1. Write the inequality as an equation and solve the equation.

2. If the inequality contains a variable in any denominator, determine the value or values that make the denominator equal to 0.

3. Construct a number line. Mark each solution that is determined in step one, and values obtained in step two on the number line. Make sure you mark these values from the lowest value on the left to the greatest value on the right.

4. Select a test value in each region of the number line.

5. Test each value in step four of the inequality to determine if it satisfies the inequality.

6. Test each boundary value to determine if it is a solution to the inequality. Remember, division by 0 is not permitted.

7. Write the solution in the form requested by your instructor.

Example 2
Solve $x^2 + 7x \geq -12$.

Solution

1. Rewrite $x^2 + 7x \geq -12$ as an equation.
$$x^2 + 7x = -12$$
$$x^2 + 7x + 12 = 0$$
$$(x+4)(x+3) = 0$$
Solving, we have $x = -4$ or $x = -3$.

2. Construct a number line:

$$
\begin{array}{c|c|c}
A & B & C \\
\hline
-4 & -3 &
\end{array}
$$

3. Select a test value in each region and determine if this value satisfies the original inequality. We shall choose −5 in region A, −3.5 in region B, and 0 in region C.

Region A: Use $x = -5$	Region B: Use $x = -3.5$	Region C: Use $x = 0$
Is $x^2 + 7x \geq -12$?	Is $x^2 + 7x \geq -12$?	Is $x^2 + 7x \geq -12$?
$(-5)^2 + 7(-5) \geq -12$	$(-3.5)^2 + 7(-3.5) \geq -12$	$0^2 + 7(0) \geq -12$
$25 + -35 \geq -12$	$-12.25 \geq -12$ False	$0 \geq -12$ True
$-10 > -12$ True		

For this problem, the boundary values are included since the sign of the original inequality is $\geq$.

4. The solution set is the values in region A along with the values in region C: $\{x | x \le -4 \text{ or } x \ge -3\}$.

Example 3

Solve $\dfrac{x-2}{x+3} \le 0$ and graph the solution set.

Solution

1. Solve $\dfrac{x-2}{x+3} = 0$. This implies that $x = 2$.

2. Determine values which make the denominator equal to zero. $x + 3 = 0$ implies that $x = -3$.

3. Construct a number line using the values found in steps 1 and 2.

A		B		C

   $-3 \qquad\qquad 2$

4. Use the test values of −4, 0, 3.

Region A: Use $x = -4$.	Region B: Use $x = 0$.	Region C: Use $x = 3$.
Is $\dfrac{x-2}{x+3} \le 0$?	Is $\dfrac{x-2}{x+3} \le 0$?	Is $\dfrac{x-2}{x+3} \le 0$?
$\dfrac{-4-2}{-4+3} \le 0$	$\dfrac{0-2}{0+3} \le 0$	$\dfrac{3-2}{3+3} \le 0$
$\dfrac{-6}{-1} \le 0$	$-\dfrac{2}{3} \le 0$ True	$\dfrac{1}{6} \le 0$ False
$6 \le 0$ False		

5. The solutions are values of region B. Since the original inequality has a $\le$ symbol, the boundary value of 2 is included. However, the value −3 is **not included since this number makes the denominator equal to zero and we cannot divide by zero.**

In solution set form, our solution is $\{x | -3 < x \le 2\}$.

Exercise Set 12.6

Solve each inequality and graph the solution on the real number line.

1. $x^2 - 9 > 0$

2. $2x^2 + 6x + 4 \le 0$

3. $(x+4)(x+2)(x-3) \ge 0$

4. $\dfrac{x-5}{x+2} < 0$

5. $\dfrac{x+3}{x} \ge 0$

6. $\dfrac{x+6}{(x-2)(x+4)} > 0$

7. $\dfrac{x-2}{x+3} < 0$

8. $\dfrac{x+4}{x-3} > 0$

9. $\dfrac{2}{2x-1} > 2$

10. $\dfrac{y}{y+4} < 1$

Answers to Exercise Set 12.6

1. $\{x|x<-3 \text{ or } x>3\}$

2. $\{x|-2 \le x \le -1\}$

3. $\{x|-4 \le x \le -2 \text{ or } x \ge 3\}$

4. $\{x|-2<x<5\}$

5. $\{x|x \le -3 \text{ or } x>0\}$

6. $\{x|-6<x<-4 \text{ or } x>2\}$

7. $\{x|-3<x<2\}$

8. $\{x|x<-4 \text{ or } x>3\}$

9. $\left\{x\left|\dfrac{1}{2}<x<1\right.\right\}$

10. $\{y|y>-4\}$

Chapter 12 Practice Test

1. Use the square root property to solve $(3x-4)^2 = 60$.

2. Solve $x^2 - 8x + 15 = 0$ by completing the square.

3. Solve $2x^2 - 8x = -64$ by completing the square.

For problems 4, 5, and 6, determine whether the equation has two distinct real solutions, a single real solution, or no real solutions.

4. $x^2 - x + 8 = 0$

5. $y^2 - 12y = -36$

6. $2x^2 + 6x + 7 = 0$

7. Solve $x^2 - 6x + 7 = 0$ using the quadratic formula.

Find the solution to the following quadratic equations by any method you choose.

8. $x^2 + 3x - 6 = 0$

9. $x^2 - x + 42 = 0$

10. $x^2 = \dfrac{5}{6}x + \dfrac{25}{6}$

Determine whether the parabola opens upward or downward, find the y-intercept, find the vertex, find the x intercepts if they exist, and sketch the graph.

11. $y = x^2 + 2x - 8$

Solve the following equations:

12. $\dfrac{5(x-1)}{x+1} = \dfrac{x-1}{x}$

13. $2y^6 - 7y^3 = -6$

Solve each inequality and express the answer in interval notation.

14. $(x - 5)(x + 2)(x + 6) \leq 0$ **15.** $\dfrac{x + 2}{x + 4} \geq 0$

Answers to Chapter 12 Practice Test

1. $\dfrac{4 \pm 2\sqrt{15}}{3}$ **2.** $3, 5$ **3.** $2 \pm 2i\sqrt{7}$

4. No real   **5.** One real   **6.** No real

7. $3 \pm \sqrt{2}$ **8.** $\dfrac{-3 \pm \sqrt{33}}{2}$ **9.** $\dfrac{1 \pm i\sqrt{167}}{2}$

10. $\dfrac{5}{2}, -\dfrac{5}{3}$

11. upward; -8; $(-1, -9)$; $-4, 2$

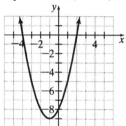

12. $\dfrac{1}{4}, 1$ **13.** $\sqrt[3]{\dfrac{3}{2}}, \sqrt[3]{2}$ **14.** $(-\infty, -6] \cup [-2, 5]$

15. $(-\infty, -4) \cup [-2, \infty)$

Chapter 13

13.1 Composite and Inverse Functions

Summary

Composition of Functions

The composition of function f with function g is denoted by $(f \circ g)(x)$ and is defined as $f[g(x)]$.

The composition of function g with function f is denoted by $(g \circ f)(x)$ and is defined as $g[f(x)]$.

Note: to find $(f \circ g)(x)$, we substitute $g(x)$ for x in $f(x)$ to get $f[g(x)]$.

Example 1

Let $f(x) = x^2 + 3x - 1$ and $g(x) = 3x - 2$. Find

    a. $(f \circ g)(x)$ **b.** $(f \circ g)(4)$

    c. $(g \circ f)(x)$ **d.** $(g \circ f)(4)$

Solution

    a. $f(x) = x^2 + 3x - 1$ so that

$$(f \circ g)(x) = f[g(x)] = (3x - 2)^2 + 3(3x - 2) - 1$$
$$= 9x^2 - 12x + 4 + 9x - 6 - 1$$
$$= 9x^2 - 3x - 3$$

    b. To find $(f \circ g)(4)$ use the above result with $x = 4$.

$$(f \circ g)(x) = 9x^2 - 3x - 3$$
$$(f \circ g)(4) = 9(4)^2 - 3(4) - 3$$
$$= 144 - 12 - 3$$
$$= 129$$

    c. $g(x) = 3x - 2$ so that

$$(g \circ f)(x) = g[f(x)] = 3(x^2 + 3x - 1) - 2$$
$$= 3x^2 + 9x - 3 - 2$$
$$= 3x^2 + 9x - 5$$

    d. To find $(g \circ f)(4)$ use the above result with $x = 4$.

$$(g \circ f)(x) = 3x^2 + 9x - 5$$
$$(g \circ f)(4) = 3(4)^2 + 9(4) - 5$$
$$= 48 + 36 - 5$$
$$= 79$$

Summary

One-to-One Functions

A **one-to-one function** is a function where each *y* value has a unique *x* value. For a function to be a one-to-one, it must pass not only a **vertical line test** (to determine whether or not it is a function) but also a **horizontal line test** (the criterion for a one-to-one function).

If $f(x)$ is a one-to-one function with ordered pairs of the form (a, b), then its inverse function, denoted by $f^{-1}(x)$, will be a one-to-one function with ordered pairs of the form (b, a).

To Find the Inverse Function of a One-to-One Function of the Form $y = f(x)$

1. Replace $f(x)$ with *y*.

2. Interchange the two variables *x* and *y*.

3. Solve the equation for *y*. The resulting equation will be the inverse function.

4. Replace *y* with $f^{-1}(x)$ using inverse notation.

Example 1
Determine which of the following are one-to-one functions:

a.

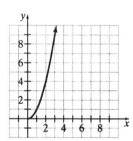

b. $\{(2, 3), (3, 9), (4, 9), (6, 10)\}$

c. $y = -2x + 5$

d. $y = -2x^2 + 1$

Solution

a. The graph of this function passes the horizontal line test so it is one-to-one.

b. The *y* value of 9 corresponds to **both 3 and 4**. Hence, this function is **not** one-to-one since each *y* value must correspond to a **unique *x* value**.

c. The graph of $y = -2x + 5$ is a straight line with slope of –2 and y-intercept of 5. The graph of **all lines which are not vertical or horizontal** will pass the horizontal line test. Thus, these lines are one-to-one and have inverses. Thus, $y = -2x + 5$ is one-to-one and has an inverse.

d. The graph of $y = -2x^2 + 1$ is a parabola opening downward. All horizontal lines will intersect this graph **twice** (except for the horizontal line passing through the vertex of the parabola). Since the graph fails the horizontal line test, $y = -2x^2 + 1$ is **not one-to-one** and does not have an inverse.

Exmaple 2
Show that $f(x) = \{(-1, 3), (1, 4), (2, 7), (3, 9)\}$ is one-to-one and find its inverse.

Solution
Each value of the range, 3, 4, 7 and 9 corresponds to a **unique value of x**. Hence, $f(x)$ is one-to-one. To find the inverse function, denoted by $f^{-1}(x)$, simply interchange the x and y values of the original function:

$f^{-1}(x) = \{(3, -1), (4, 1), (7, 2), (9, 3)\}$.

Notice that the domain of $f^{-1}(x)$ is the same as the range of $f(x)$ and vice versa. **This will always be the case for functions and their inverses.**

Example 3

a. Find $f^{-1}(x)$ if $f(x) = -2x + 5$.

b. Graph $f^{-1}(x)$ and $f(x)$ on the same graph.

c. Find $(f \circ f^{-1})(x)$.

Solution

a. 1. $y = -2x + 5$ upon replacing $f(x)$ with y.

   2. Interchange x and y: $x = -2y + 5$

   3. Solve for y: $x = -2y + 5$
   $$2y = 5 - x$$
   $$y = \frac{5 - x}{2}$$

   4. or $f^{-1}(x) = \frac{5 - x}{2}$

b.

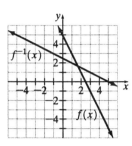

c. $f(x) = -2x + 5$ so that

$$(f \circ f^{-1})(x) = -2\left(\frac{5-x}{2}\right) + 5$$
$$= -(5-x) + 5$$
$$= -5 + x + 5$$
$$= x, \text{ as expected}$$

Exercise Set 13.1

1. Let $f(x) = x^2 + 2$ and $g(x) = x^2 - 4x + 5$. Find

   a. $(f \circ g)(x)$ **b.** $(f \circ g)(1)$

   c. $(g \circ f)(x)$ **d.** $(g \circ f)(1)$

Determine if the functions below are one-to-one.

2.   **3.**   **4.**

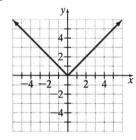 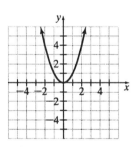 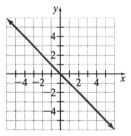

5. $\{(1, 3), (2, 5), (3, 7), (4, 3), (-2, 9)\}$

6. $y = x^2 - 9, \; x \geq 0$ **7.** $y = |x|, \; x \leq 0$

Find $f^{-1}(x)$ (problems 8, 9, and 10)

8. $f(x) = x - 3$ **9.** $f(x) = 3x + 5$ **10.** $f(x) = -3x + 7$

11. Show that $(f \circ f^{-1})(x) = x$ for $f(x) = 3x + 2$ and $f^{-1}(x) = \dfrac{x-2}{3}$.

12. Show that $(f^{-1} \circ f)(x) = x$ for $f(x) = x^3 + 1$ and $f^{-1}(x) = \sqrt[3]{x-1}$.

Answers to Exercise Set 13.1

1. a. $x^4 - 8x^3 + 26x^2 - 40x + 27$

    b. 6   **c.** $x^4 + 1$ **d.** 2

2. No   **3.** No   **4.** Yes

5. No   **6.** Yes   **7.** Yes

8. $f^{-1}(x) = x + 3$ **9.** $f^{-1}(x) = \dfrac{x-5}{3}$ **10.** $f^{-1}(x) = \dfrac{7-x}{3}$

11. $(f \circ f^{-1})(x) = f[f^{-1}(x)] = 3\left(\dfrac{x-2}{3}\right) + 2 = x - 2 + 2 = x$

12. $(f^{-1} \circ f)(x) = f^{-1}[f(x)] = \sqrt[3]{(x^3 + 1) - 1} = \sqrt[3]{x^3} = x$

13.2 Exponential Functions

Summary

> An **exponential equation** or function is one that has a variable as an exponent.

Example 1
A biologist has one cell at the start of an experiment. The number of cells doubles each day. The formula for the number of cells after x days is given by $y = 2^x$. Find the number of cells present after 20 days.

Solution
$y = 2^x$
If $x = 20$, then $y = 2^{20} = 1,048,576$ (calculator).

This example illustrates how fast the exponential function increases.

Summary

> **Exponential Function**
>
> For any real number $a > 0$ and $a \neq 1$, $f(x) = a^x$ is an exponential function.

Example 2
Graph $f(x) = 3^x$ and state the domain, range and y-intercept.

Solution
Find several ordered pairs of the function by replacing x with arbitrary values.

Ordered pair

If $x = -3$, then $y = 3^{-3} = \dfrac{1}{3^3} = \dfrac{1}{27}$ $\qquad \left(-3, \dfrac{1}{27}\right)$

If $x = -1$, then $y = 3^{-1} = \dfrac{1}{3}$ $\qquad \left(-1, \dfrac{1}{3}\right)$

If $x = 0$, then $y = 3^0 = 1$ $\qquad (0, 1)$

If $x = 1$, then $y = 3^1 = 3$ $\qquad (1, 3)$

If $x = 2$, then $y = 3^2 = 9$ $\qquad (2, 9)$

If $x = 3$, then $y = 3^3 = 27$ $\qquad (3, 27)$

Plot the ordered pairs and connect them with a smooth curve.

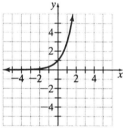

The domain is the set of real numbers: $\{x | x \text{ is a real number}\}$. The range is $\{y | y > 0\}$. The y-intercept is (0, 1).

Example 3
The formula $A = p(1+r)^n$ gives the amount, A, in an account in which p dollars is compounded annually for n years at an interest rate of r. If \$5000 is invested at 6% compounded annually for 10 years, how much will be in the account at the end of 10 years?

Solution
Use $A = p(1+r)^n$ with $p = 5000$, $r = 0.06$, $n = 1$, and $t = 10$.
Then, $A = 5000(1 + 0.06)^{10} = 5000(1.06)^{10} \approx 5000(1.790847)$
$\qquad A \approx \$8,954.24$

Example 4
In 1998, the population in Xenobia was 103,000 people and is growing according to the function
$P(t) = 103,000(1.08)^t$ where t is the number of years after 1998. Estimate the population in 2008.

Solution
Let $t = 10$, the number of years after 1998.
Then $P(10) = 103,000(1.08)^{10}$
$$\approx 103,000(2.158925)$$
$$\approx 222,369$$

Example 5
Bob invests $100 into a savings account earning interest at 6% compounded semiannually. Find the amount in the account after 20 years.

Solution

Use $A = p\left(1 + \dfrac{r}{n}\right)^{nt}$ with $p = 100$, $n = 2$, $r = 0.06$, and $t = 20$.

Then, $A = 100\left(1 + \dfrac{0.06}{2}\right)^{2 \cdot 20}$

$$= 100(1.03)^{40}$$
$$\approx 100(3.262038)$$
$$\approx 326.20$$

At the end of 20 years, Bob will have $326.20 in the account.

Exercise Set 13.2

Graph the exponential functions.

1. $y = 4^x$
2. $y = \left(\dfrac{1}{4}\right)^x$
3. $y = 2^{-x}$

4. If $5,000 is invested at 6% compounded annually, find the amount in the account at the end of

   a. 4 years
   b. 10 years
   c. 20 years

5. In the formula $N(x) = 2^x$, N is the number of one-celled organisms after x hours.

   a. How many one-celled organisms are there after 13 hours?

   b. In how many hours will 1024 organisms be present?

Answers for Exercise Set 13.2

1.

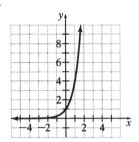

2.

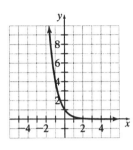

3.

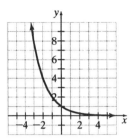

4. a. $6312.38

b. $8,954.24

c. $16,035.68

5. a. 8192

b. 10 hours

13.3 Logarithmic Functions

Summary

Logarithms

For all positive numbers a, where $a \neq 1$, $y = \log_a x$
means $a^y = x$.

Example 1
Write in logarithmic form.

a. $10^2 = 100$

b. $\left(\dfrac{1}{3}\right)^4 = \dfrac{1}{81}$

c. $5^{-3} = \dfrac{1}{125}$

Solution

a. $10^2 = 100$ is equivalent to $\log_{10} 100 = 2$.

b. $\left(\dfrac{1}{3}\right)^4 = \dfrac{1}{81}$ is equivalent to $\log_{1/3}\left(\dfrac{1}{81}\right) = 4$

c. $5^{-3} = \dfrac{1}{125}$ is equivalent to $\log_5\left(\dfrac{1}{125}\right) = -3$

Example 2
Write in exponential form and then find the missing value.

a. $y = \log_6 36$

b. $\log_{10} x = -1$

c. $\log_b 4 = -2$

Solution

a. $y = \log_6 36$ is equivalent to $6^y = 36$. Thus, $y = 2$.

b. $\log_{10} x = -1$ is equivalent to $10^{-1} = x$. Thus, $x = \dfrac{1}{10}$ or 0.1.

c. $\log_b 4 = -2$ is equivalent to $b^{-2} = 4$ or $\dfrac{1}{b^2} = 4$.

Solve this equation to obtain $b^2 = \dfrac{1}{4}$ or $b = \pm\sqrt{\dfrac{1}{4}} = \pm\dfrac{1}{2}$.

Since b, the base, must be a positive number, use $b = \dfrac{1}{2}$.

Example 3
Graph $y = \log_3 x$ and state the domain and range.

Solution

$y = \log_3 x$ means $3^y = x$. It is preferable to choose the y values first and then find the corresponding x values.

x	$\frac{1}{27}$	$\frac{1}{9}$	$\frac{1}{3}$	1	3	9	27
y	-3	-2	-1	0	1	2	3

Connect the points with a smooth curve.

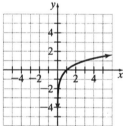

The domain is $(0, \infty)$ and the range is $(-\infty, \infty)$.

Since $y = \log_3 x$ means $3^y = x$, this function is the inverse of the function $y = 3^x$. The graphs of $y = \log_3 x$ and $y = 3^x$ are symmetric with respect to the line $y = x$.

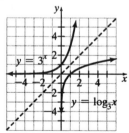

$y = x$

Exercise Set 13.3

Write in logarithmic form.

1. $5^4 = 625$

2. $2^{-5} = \dfrac{1}{32}$

3. $a^y = b$

Write in exponential form.

4. $\log_5 25 = 2$

5. $\log_6\left(\dfrac{1}{36}\right) = -2$

6. $\log_{1/2} y = x$

Find the value of each unknown.

7. $\log_7 1 = y$

8. $\log_b\left(\dfrac{1}{4}\right) = 2$

9. $\log_2 x = 3$

Answers for Exercise Set 13.3

1. $\log_5 625 = 4$

2. $\log_2\left(\dfrac{1}{32}\right) = -5$

3. $\log_a b = y$

4. $5^2 = 25$

5. $6^{-2} = \dfrac{1}{36}$

6. $\left(\dfrac{1}{2}\right)^x = y$

7. $y = 0$

8. $b = \dfrac{1}{2}$

9. $x = 8$

13.4 Properties of Logarithms

Summary

Product Rule for Logarithms

For positive real numbers x, y, and a, $a \neq 1$,

$$\log_a xy = \log_a x + \log_a y$$

Quotient Rule for Logarithms

$$\log_a \frac{x}{y} = \log_a x - \log_a y$$

Power Rule for Logarithms

If x and a are positive real numbers, $a \neq 1$, and n is any real number, then

$$\log_a x^n = n \log_a x$$

Additional Properties

If $a > 0$, $a \neq 1$, then $\log_a a^x = x$ and $a^{\log_a(x)} = x$

Example 1
Expand the following.

    a. $\log_6(7 \cdot 9)$ **b.** $\log_{10} y(y-3)$

Solution

    a. $\log_6(7 \cdot 9) = \log_6 7 + \log_6 9$ **b.** $\log_{10} y(y-3) = \log_{10} y + \log_{10} y - 3$

Example 2
Expand the following.

    a. $\log_5\left(\dfrac{17}{8}\right)$ **b.** $\log_3 \sqrt{\dfrac{x}{x-2}}$

Solution

    a. $\log_5\left(\dfrac{17}{8}\right) = \log_5 17 - \log_5 8$

b. $\log_3 \sqrt{\dfrac{x}{x-2}} = \log_3\left(\dfrac{x}{x-2}\right)^{1/2}$

$\qquad\qquad = \dfrac{1}{2}\log_3\left(\dfrac{x}{x-2}\right)$

$\qquad\qquad = \dfrac{1}{2}[\log_3 x - \log_3(x-2)]$

Example 3
Write as a logarithm of a single expression.

a. $3\log_a x + 2\log_a y$

b. $\dfrac{1}{2}[\log_5(2-y) - \log_5 7]$

Solution

a. $3\log_a x + 2\log_a y = \log_a x^3 + \log_a y^2 = \log_a(x^3 y^2)$

b. $\dfrac{1}{2}[\log_5(2-y) - \log_5 7] = \dfrac{1}{2}\log_5\left(\dfrac{2-y}{7}\right) = \log_5\left(\dfrac{2-y}{7}\right)^{1/2}$ or $\log_5\sqrt{\dfrac{2-y}{7}}$

Example 4
Evaluate

a. $7^{\log_7(13)}$

b. $(3^2)^{\log_9 5}$

c. $2\log_9\left(\sqrt{9}\right)$

Solution

a. $7^{\log_7(13)}$ is of the form $a^{\log_a(x)}$. Since $a^{\log_a(x)} = x$, $7^{\log_7(13)} = 13$.

b. $(3^2)^{\log_9 5}$ is equivalent to $9^{\log_9(5)} = 5$.

c. $2\log_9 \sqrt{9} = \log_9\left(\sqrt{9}\right)^2 = \log_9 9 = 1$

Exercise Set 13.4

Expand the following.

1. $\log_9(8 \cdot 5)$

2. $\log_7 x(x-5)$

3. $\log_{10}\dfrac{\sqrt{y}}{y-3}$

Write as a logarithm of a single expression.

4. $3\log_5 x + 2\log_5(x-1)$

5. $\log_7 6 + 4\log_7 x$

6. $\dfrac{1}{2}[\log_6(x-1) - \log_6 3]$

Evaluate.

7. $\log_3 3^5$

8. $\log_{12} 1$

9. $(5^3)^{\log_{125} 1}$

Answers to Exercise Set 13.4

1. $\log_9 8 + \log_9 5$

2. $\log_7(x) + \log_7(x - 5)$

3. $\dfrac{1}{2} \log_{10} y - \log_{10}(y - 3)$

4. $\log_5[x^3 \cdot (x - 1)^2]$

5. $\log_7(6x^4)$

6. $\log_6 \sqrt{\dfrac{x - 1}{3}}$

7. 5

8. 0

9. 1

13.5   Common Logarithms

Summary

Common Logarithms

The **common logarithm** of a positive real number is the **exponent** to which the base **10** is raised to obtain the number.

If $\log N = L$, then $10^L = N$

Antilogarithms

If $\log N = L$ then $N = $ antilog L.

Example 1
Find the following using a scientific or graphing calculator.

a. log 769

b. log 1

c. log 0.000123

d. log (–5.9)

Solution

a. log 769 = 2.8859

b. log 1 = 0 (This problem can be done without using a calculator. Log 1 = 0 since $10^0 = 1$.)

c. log 0.000123 = –3.9101

d. log (–5.9) does not exist since the domain of the logarithm function is $\{x | x > 0\}$.

Example 2
From example 1, it was found that log 769 = 2.8859. This means that $10^{2.8859} \approx 769$. We say the **antilogarithm** of 2.8859 is 769 since $10^{2.8859} = 769$.

Example 3
Find N if log $N = 1.234$.

Solution
If log $N = 1.234$ then $10^{1.234} = N$ or 17.14.

Exercise Set 13.5

Find the common logarithm of the number and round answers to 4 decimal places.

    1. 983   **2.** 7   **3.** 29,541

    4. 0.00543

Find the antilog of the logarithm. Round answers to 3 decimal places.

    5. 2.123   **6.** 6.49   **7.** −2.965

Find N. Round N to 3 decimal places.

    8. log $N = 2.497$ **9.** log $N = -2.143$ **10.** log $N = 3$

Find the common logarithms without using a calculator.

    11. log 1000   **12.** log 0.001   **13.** $\log 10^7$

    14. $10^{\log 50}$ **15.** $6 \cdot \log 10^{5.2}$

Answers to Exercise Set 13.5

    1. 2.9926   **2.** .8451   **3.** 4.4704

    4. −2.2652   **5.** 132.739   **6.** 3,090,295.433

    7. 0.001   **8.** 314.051   **9.** 0.007

   10. 1000   **11.** 3   **12.** −3

   13. 7   **14.** 50   **15.** 31.2

13.6 Exponential and Logarithmic Equations

Summary

**Properties for Solving Exponential and
Logarithmic Equations**

a. If $x = y$, then $a^x = a^y$.

b. If $a^x = a^y$, then $x = y$.

c. If $x = y$, then $\log x = \log y$. $(x > 0, y > 0)$

d. If $\log x = \log y$, then $x = y$ $(x > 0, y > 0)$

Example 1
Solve $3^{-x} = 27$.

Solution
$3^{-x} = 27$
$3^{-x} = 3^3$
$-x = 3$ (If $a^x = a^y$, then $x = y$)
$\quad x = -3$

Example 2
Solve $3^x = 6$.

Solution
$$3^x = 6$$
$\log 3^x = \log 6$ (if $x = y$, then $\log x = \log y$)
$x \log 3 = \log 6$ (property of logarithms)
$$x = \frac{\log 6}{\log 3} = 1.631$$

Note: The numerical value was not calculated until the very last step. The reason is to minimize accumulated round off errors. For example, if log 6 was evaluated in the first step, some students would round off the answer differently. Again, in step 2, log 3 might be rounded differently by different students. The quotient of these two rounded numbers would vary depending upon the rules used for rounding log 3 and log 6. The answer might also vary depending upon the number of decimal places your calculator rounds to.

Example 3

Solve $5^x = 3^{x+1}$.

Solution

$$5^x = 3^{x+1}$$
$$\log 5^x = \log 3^{x+1} \quad (\text{if } x = y, \text{ the } \log x = \log y)$$
$$x \log 5 = (x+1) \log 3$$
$$x \log 5 = x \log 3 + \log 3$$
$$x \log 5 - x \log 3 = \log 3$$
$$x(\log 5 - \log 3) = \log 3$$
$$x = \frac{\log 3}{\log 5 - \log 3} = 2.151$$

Example 4

Solve $\log_2(x-2) = 3 - \log_2 x$.

Solution

$$\log_2(x-2) = 3 - \log_2 x$$
$$\log_2(x-2) + \log_2 x = 3$$
$$\log_2[(x-2)x] = 3$$
$$2^3 = (x-2)x$$
$$8 = x^2 - 2x$$
$$0 = x^2 - 2x - 8$$
$$0 = (x-4)(x+2)$$
$$x = 4, \ x = -2$$

Discard the solution of $x = -2$ since the domain of $\log_2 x$ must consist of numbers greater than zero. The solution is $x = 4$.

Example 5

The formula $A = p(1+r)^n$ gives the amount, A, to which a principal, p, will grow if compounded annually at an interest rate of r for n compounding periods. How long would it take \$1000 to grow to \$2000 if compounded annually at 6% interest?

Solution

Use $A = p(1+r)^n$, where $A = 2000$, $p = 1000$, $r = 0.06$ and n is unknown.

$$2000 = 1000(1 + 0.06)^n$$
$$2 = (1.06)^n$$
$$\log 2 = \log(1.06)^n$$
$$\log 2 = n \log(1.06)$$
$$\frac{\log 2}{\log 1.06} = n$$
$$11.89 = n$$

To the nearest year, $n = 12$ years.

Exercise Set 13.6

Solve the equation without using a calculator.

1. $2^{-x} = 16$

2. $5^{-x} = \dfrac{1}{125}$

3. $\left(\dfrac{1}{2}\right)^x = 16$

4. $64^x = 4^{2x-1}$

Use a calculator to solve the following equations.

5. $5^x = 75$

6. $6^x = 23$

7. $\left(\dfrac{1}{4}\right)^x = 9$

8. $5^x = 3^{x+2}$

Solve.

9. $\log_3(x+6) = \log_3(3x-2)$

10. $\log_3(x-2) = 1 - \log_3(x)$

11. $\log_7(5x+4) = 2$

12. If the initial amount of bacteria in a culture is 4500, when will the number of bacteria in this culture reach 50,000? Use the formula $N = 4500 \cdot 2^t$, where t is in hours and N represents the number of bacteria.

Answers to Exercise Set 13.6

1. –4

2. 3

3. –4

4. –1

5. 2.683

6. 1.750

7. –1.585

8. 4.301

9. 4

10. 3

11. 9

12. 3.47 hours

13.7   Natural Exponential and Natural Logarithmic Functions

Summary

Natural Exponential Function

The Natural Exponential function is

$$f(x) = e^x, \text{ where } e \approx 2.7183.$$

Natural Logarithms

Natural logarithms are logarithms to the base e. Natural logarithms are indicated by the letters ln.

$$\log_e(x) = \ln x$$

If $y = \ln x$ then $e^y = x$.

The functions $y = e^x$ and $y = \ln(x)$ are inverse functions.

Example 1
Use a calculator to evaluate each of the following.

    a. ln 3.1   **b.** ln 0.00013

    c. $e^{7.314}$ **d.** e^{-2}

Solution

    a. ln 3.1 = 1.131 to three decimal places.

    b. ln 0.00013 = −8.948

    c. $e^{7.314} = 1501.17$

    d. $e^{-2} = 0.135$

Example 2
Find N if

    a. ln $N = 3.48$

    b. ln $N = -0.037$

Solution

    a. If ln $N = 3.48$, then $e^{3.48} = N = 32.46$

b. If $\ln N = -0.037$, then $e^{-0.037} = N = 0.964$

Summary

Change of Base Formula

If $a > 0$, $b > 0$, $x > 0$ and neither a nor b equals 1, then

$$\log_a x = \frac{\log_b x}{\log_b a}.$$

Summary

Properties for Natural Logarithms

$$\ln xy = \ln x + \ln y, \ (x > 0 \text{ and } y > 0)$$

$$\ln \frac{x}{y} = \ln x - \ln y, \ (x > 0 \text{ and } y > 0)$$

$$\ln x^n = n \ln x, \ (x > 0)$$

Additional Properties

$$\ln e^x = x$$
$$e^{\ln x} = x, \ (x > 0)$$

Example 3
Find $\log_8 5$.

Solution
Use $\log_a x = \dfrac{\log_b x}{\log_b a}$.

$$\log_8 5 = \frac{\log 5}{\log 8} \approx 0.77398$$

Check: If $\log_8 5 = .77398$, then $8^{0.77398} = 5$ (which is true).

An alternate method of finding $\log_8 5$ would be to use the change of base formula with natural logarithms, rather than common logarithms: $\log_8 5 = \dfrac{\ln 5}{\ln 8}$.

Example 4
Solve the equation $\ln x - \dfrac{3}{2} \ln 4 = 0$.

Solution

$$\ln x - \frac{3}{2}\ln 4 = 0$$

$$\ln x = \frac{3}{2}\ln 4$$

$$\ln x = \ln 4^{3/2}$$

This implies that $x = 4^{3/2} = \left(\sqrt{4}\right)^3 = 2^3 = 8.$

Example 5

Solve $e^{-0.12t} = 0.8.$

Solution

$$e^{-0.12t} = 0.8$$

$$\ln e^{-0.12t} = \ln 0.8$$

$$-0.12t \ln e = \ln 0.8$$

$$-0.12t(1) = \ln 0.8$$

$$t = \frac{\ln 0.8}{-0.12} = 1.86$$

Example 6

How much money must be deposited today to become $15,000 in 12 years if invested at 7% compounded continuously?

Solution

Use the formula $A = pe^{rt}$ with $A = 15,000$, $r = 0.07$ and $t = 12$.

$$15000 = pe^{(0.07)(12)}$$

$$\frac{15,000}{e^{(0.07)(12)}} = p$$

$$6475.66 = p$$

Exercise Set 13.7

Find the following values correct to 3 decimal places.

1. $\ln 397$ **2.** $\ln 0.00395$

Find the value of N rounded to 3 significant digits.

3. $\ln N = 2.9$ **4.** $\ln N = -4.75$

Use the change of base formula to evaluate the following.

5. $\log_7 9$ **6.** $\log_5 0.796$

Solve for the variable.

7. $\ln x - \ln 2 + \ln(x - 1) = 0$ **8.** $\ln 8 - \ln x = 5 \cdot \ln 2$

9. $20 = 40 \cdot e^{-0.5t}$

10. $180 = 90 \cdot e^{3t}$

11. How much money must be deposited today to become $40,000 in 16 years if invested at 5% compounded continuously?

12. The atmospheric pressure, P, in pounds per square inch, at an elevation of x feet above sea level can be found by the formula $P = 14.7 \cdot e^{-0.0004x}$. Find the atmospheric pressure at an elevation of 1 mile (5,280 feet).

Answers to Exercise Set 13.7

1. 5.984

2. –5.534

3. 18.174

4. 0.009

5. 1.129

6. –0.142

7. $x = 2$

8. $\dfrac{1}{4} = x$

9. $t = 1.386$

10. $t = 0.231$

11. $p = \$17,973.16$

12. Pressure is 1.78 pounds

Chapter 13 Practice Test

Graph the following functions.

1. $y = 2^{-x}$

2. $y = \log_{1/2}(x)$

Write in logarithmic form.

3. $6^3 = 216$

4. $16^{1/4} = 2$

Write in exponential form.

5. $\log_7 2401 = 4$

6. $\log_8 2 = \dfrac{1}{3}$

Use the properties of logarithms to expand each expression below.

7. $\log_4(5 \cdot 6)$

8. $\log_3\left(\dfrac{5}{8}\right)$

9. $\log_7 5^4$

10. $\log_6\left(\dfrac{x}{xy}\right)$

Solve for the variable.

11. $3 = \log_4 x$

12. $\log x = 2.3304$

13. $10^{\log_{10} 3} = x$

14. $27^x = 3^{2x+5}$

Evaluate.

15. a. $\log 7.96$ **b.** antilog of 2.347

Solve for the variable.

16. $\log_3 x + \log_3(2x + 1) = 1$ **17.** $4^x = 37$

18. If \$10,000 is placed in an account paying 7% interest compounded continuously, find the time needed for the account to double in value.

19. Find the inverse of $f(x) = 2x - 7$.

20. Let $f(x) = x^2 + 10x - 5$ and $g(x) = 2x - 1$. Find $(f \circ g)(x)$ and $(g \circ f)(x)$.

Answers to Chapter 13 Practice Test

1.

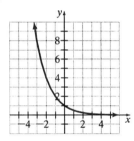

2.

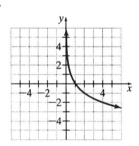

3. $\log_6 216 = 3$

4. $\log_{16} 2 = \dfrac{1}{4}$ **5.** $7^4 = 2410$ **6.** $8^{1/3} = 2$

7. $\log_4 5 + \log_4 6$ **8.** $\log_3 5 - \log_3 8$ **9.** $4\log_7 5$

10. $-\log_6 y$ **11.** $x = 64$ **12.** $x = 214$

13. $x = 3$ **14.** $x = 5$

15. a. 0.9009 **b.** 222.331

16. $x = 1$ **17.** $x = 2.605$ **18.** 9.90 years

19. $f^{-1}(x) = \dfrac{x + 7}{2}$ **20.** $(f \circ g)(x) = 4x^2 + 16x - 14$

                                            $(g \circ f)(x) = 2x^2 + 20x - 11$

Chapter 14

14.1 The Parabola and the Circle

Summary

Parabola with Vertex at (h, k)

1. $y = a(x - h)^2 + k$, $a > 0$
   (opens upward)

2. $y = a(x - h)^2 + k$, $a < 0$
   (opens downward)

3. $x = a(y - k)^2 + h$, $a > 0$
   (opens to the right)

4. $x = a(y - k)^2 + h$, $a < 0$
   (opens to the left)

Example 1

Sketch the graph of $y = 2(x + 1)^2 - 3$.

Solution
This is of the form of equation 1 $(a = 2 > 0)$.
Therefore, vertex is at $(-1, -3)$ and it opens upward.

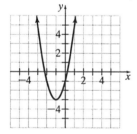

Example 2

Sketch the graph of $y = -2x^2 + 4x - 3$.

Solution
First, factor -2 from the from the two terms containing the variable to make the coefficient of the squared term equal to 1.

$y = -2(x^2 - 2x) - 3$

Now add $\left(\dfrac{2}{2}\right)^2 = (1)^2 = 1$ to complete the square. Since the added 1 must be multiplied by –2, we are really adding $(-2)(1) = -2$.

Therefore, add +2 to the –3 to keep the equation the same.

$y = \underline{-2}(x^2 - 2x + \underline{1}) - 3 + \underline{2}$

$y = -2(x - 1)^2 - 1$

The parabola opens downward since $a = -2 < 0$ and the vertex is at $(1, -1)$.

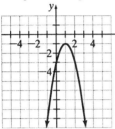

Example 3

Sketch the graph of $x = y^2 - 4y + 1$.

Solution

First we complete the square for $y^2 - 4y$ by adding $\left(\dfrac{4}{2}\right)^2 = 4$. We must add –4 to 1 to keep the equation the same.

$x = (y^2 - 4y + 4) + 1 - 4$

$x = (y - 2)^2 - 3$

This is of the form of equation 3, $a = 1 > 0$. Therefore, it opens to the right and the vertex is at $(-3, 2)$.

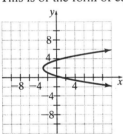

Summary

Circle with its Center at the Origin and Radius *r*
$x^2 + y^2 = r^2$

Example 4
Write the equation of a circle with center at $(0, 0)$ and radius 4.

Solution
The radius 4 should be substituted into the equaton for r.
$$x^2 + y^2 = r^2$$
$$x^2 + y^2 = 4^2$$
$$x^2 + y^2 = 16$$

Example 5
Sketch a graph of the equation $x^2 + y^2 = 25$.

Solution
Write $x^2 + y^2 = 25$
$$x^2 + y^2 = 5^2$$
It is a circle with center at the origin and radius 5.

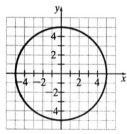

Summary

Circle with its Center at (h, k) with Radius r

$$(x - h)^2 + (y - k)^2 = r^2$$

Example 6
Write the equation of the circle with center at $(2, -1)$ and radius 3.

Solution
Substitute $h = 2$, $k = -1$, and $r = 3$ into the equation.
$$(x - 2)^2 + (y - (-1))^2 = 3^2$$
$$(x - 2)^2 + (y + 1)^2 = 9$$

Example 7

Sketch a graph of the equation of $x^2 - 4x + y^2 + 2y - 11 = 0$.

Solution

Since the coefficients of x^2 and y^2 are equal (in this case, 1), we know we have the equation of a circle. Rewrite the equation by keeping the variables on the left side of the equation and moving the constant to the right side.

$x^2 - 4x + y^2 + 2y = 11$

Complete the square for each variable.

$x^2 - 4x + \underline{4} + y^2 + 2y + \underline{1} = 11 + \underline{4} + \underline{1}$

$$(x - 2)^2 + (y + 1)^2 = 16$$

$$(x - 2)^2 + (y - (-1))^2 = 4^2$$

So the center is at $(2, -1)$ and the radius is 4.

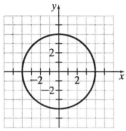

Exercise Set 14.1

Sketch the graph of each equation.

1. $y = -(x - 2)^2 + 4$ 2. $y = 3(x + 1)^2 + 2$

3. $x = (y + 3)^2 - 1$ 4. $x = -2(y - 2)^2 + 5$

Write each equation in the form $y = a(x - h)^2 + k$ or $x = a(y - k)^2 + h$ and then sketch the graph of the equation.

5. $y = -2x^2 + 4x$ 6. $y = x^2 - 4x + 2$

7. $x = -2y^2 + 12y - 13$ 8. $x = y^2 + 2y$

Write the equation of the circle with the given center and radius.

9. Center $(0, 0)$; radius 6   10. Center $(0, 0)$; radius $\sqrt{3}$

11. Center $(3, 5)$; radius 5   12. Center $(-2, 5)$; radius $\sqrt{10}$

Write the equation of the circle in standard form. Determine its center and radius. Then sketch a graph.

13. $x^2 + y^2 - 4 = 0$ **14.** $2x^2 + 2y^2 - 2 = 0$

15. $x^2 + y^2 + 2x + 4y - 4 = 0$ **16.** $x^2 + y^2 - 8x - 6y = 0$

17. $x^2 + y^2 + 12x - 2y + 1 = 0$ **18.** $x^2 + y^2 - 10x + 2y + 10 = 0$

Answers to Exercise Set 14.1

1.

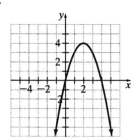

2.

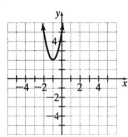

3.

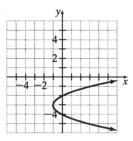

4.

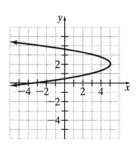

5. $y = -2(x-1)^2 + 2$

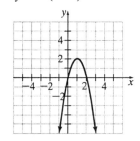

6. $y = (x-2)^2 - 2$

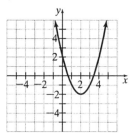

7. $x = -2(y-3)^2 + 5$

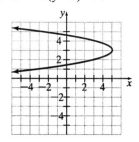

8. $x = (y+1)^2 - 1$

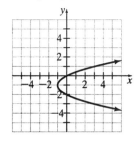

9. $x^2 + y^2 = 36$

10. $x^2 + y^2 = 3$ **11.** $(x-3)^2 + (y-5)^2 = 25$ **12.** $(x+2)^2 + (y-5)^2 = 10$

13. $x^2 + y^2 = 4$; center $(0, 0)$; radius 2

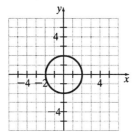

14. $x^2 + y^2 = 1$; center $(0, 0)$; radius 1

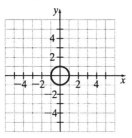

15. $(x + 1)^2 + (y + 2)^2 = 9$; center $(-1, -2)$; radius 3

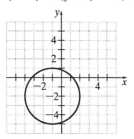

16. $(x - 4)^2 + (y - 3)^2 = 25$; center $(4, 3)$; radius 5

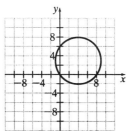

17. $(x+6)^2 + (y-1)^2 = 36$; center $(-6, 1)$; radius 6

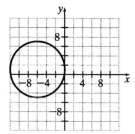

18. $(x-5)^2 + (y+1)^2 = 16$; center $(5, -1)$; radius 4

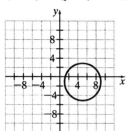

14.2 The Ellipse

Summary

Ellipse with Its Center at the Origin

$\dfrac{x^2}{a^2} + \dfrac{y^2}{b^2} = 1$, where $(a, 0)$ and $(-a, 0)$ are the

x-intercepts and $(0, b)$ and $(0, -b)$ are the y-intercepts.

Example 1

Sketch the graph of $\dfrac{x^2}{16} + \dfrac{y^2}{9} = 1$.

Solution

Rewrite $\dfrac{x^2}{16} + \dfrac{y^2}{19} = 1$ as $\dfrac{x^2}{4^2} + \dfrac{y^2}{3^2} = 1$.

The x-intercepts are −4 and 4. The y-intercepts are −3 and 3.

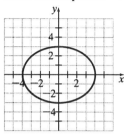

Example 2

Sketch the graph of $9x^2 + 4y^2 = 36$.

Solution

Divide each side of the equation by 36.

$$\dfrac{9x^2}{36} + \dfrac{4y^2}{36} = \dfrac{36}{36} = \dfrac{x^2}{4} + \dfrac{y^2}{9} = 1 = \dfrac{x^2}{2^2} + \dfrac{y^2}{3^2} = 1$$

The x-intercepts are −2 and 2. The y-intercepts are −3 and 3.

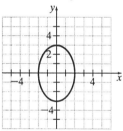

Exercise Set 14.2

Sketch the graph of each equation.

1. $\dfrac{x^2}{4} + \dfrac{y^2}{25} = 1$

2. $\dfrac{x^2}{9} + \dfrac{y^2}{50} = 1$

3. $2x^2 + 8y^2 = 32$

4. $4x^2 + 25y^2 = 100$

5. $4x^2 + y^2 = 1$

Answers to Exercise Set 14.2

1.

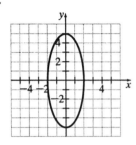

2.

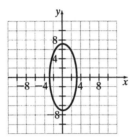

3.

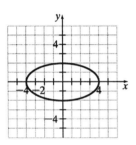

4.

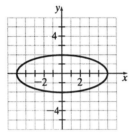

5.

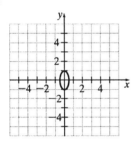

14.3 The Hyperbola

Summary

Hyperbola with its Center at the Origin

Transverse axis along x-axis (opens to right and left)

$$\frac{x^2}{a^2} - \frac{y^2}{b^2} = 1$$

Transverse axis along y-axis (opens upward and downward)

$$\frac{y^2}{b^2} - \frac{x^2}{a^2} = 1$$

Asymptotes

$$y = \frac{b}{a}x \text{ and } y = -\frac{b}{a}x$$

This may be given as $y = \pm\frac{b}{a}x$.

Example 1

Consider the equation $\frac{x^2}{4} - \frac{y^2}{9} = 1$.

   a. Determine the equation of the asymptotes of the hyperbola.

   b. Sketch the hyperbola using the asymptotes.

Solution

   a. Here, $a^2 = 4$ and $b^2 = 9$, so $a = 2$ and $b = 3$.

      Using $y = \frac{b}{a}x$ and $y = -\frac{b}{a}x$, we have $y = \frac{3}{2}x$ and $y = -\frac{3}{2}x$.

   b. First graph the asymptotes. Since the x term is positive, the transverse axis is along the x-axis. The denominator of the x term is positive, thus the x-intercepts are $\pm\sqrt{4}$ or ± 2.

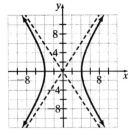

Example
Consider $4y^2 - 16x^2 = 64$.

   a. Show that the equation is a hyperbola by expressing the equation in standard form.

   b. Determine the asymptotes and sketch a graph.

Solution

   a. Divide both sides by 64 to obtain 1 on the right side:

$$\frac{4y^2}{64} - \frac{16x^2}{64} = \frac{64}{64}$$

$$\frac{y^2}{16} - \frac{x^2}{4} = 1$$

   b. Here $b^2 = 16$ so $b = 4$ and $a^2 = 4$ so $a = 2$. The asymptotes are $y = \pm\dfrac{4}{2}x = \pm 2x$.

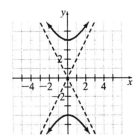

Summary

Another form of the hyperbola is $xy = c$, where c is a nonzero constant.

Example 3
Sketch the graph of $xy = 1$.

Solution

$xy = 1$ or $y = \dfrac{1}{x}$. We will make a table with x values less than zero and a similar table with x values greater than zero.

x	y		x	y
-4	$-\frac{1}{4}$		4	$\frac{1}{4}$
-2	$-\frac{1}{2}$		2	$\frac{1}{2}$
-1	-1		1	1
$-\frac{1}{2}$	-2		$\frac{1}{2}$	2
$-\frac{1}{4}$	-4		$\frac{1}{4}$	4
$-\frac{1}{8}$	-8		$\frac{1}{8}$	8

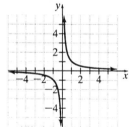

Exercise Set 14.3

Determine the equations of the asymptotes and sketch a graph of the equation.

1. $\dfrac{x^2}{4} - \dfrac{y^2}{36} = 1$

2. $\dfrac{x^2}{25} - \dfrac{y^2}{25} = 1$

3. $\dfrac{y^2}{36} - \dfrac{x^2}{49} = 1$

4. $\dfrac{y^2}{16} - \dfrac{x^2}{1} = 1$

Write each equation in standard form, determine the equation of the asymptotes, and then sketch the graph.

5. $25x^2 - 4y^2 = 100$

6. $9x^2 - 4y^2 = 36$

7. $4x^2 - y^2 = 4$

8. $10y^2 - 90x^2 = 90$

Sketch a graph of the equation.

9. $xy = 8$

10. $y = -\dfrac{2}{x}$

Answers to Exercise Set 14.3

1. $y = \pm 3x$

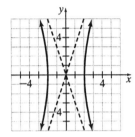

2. $y = \pm x$

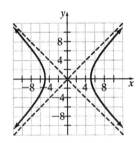

3. $y = \pm\dfrac{6}{7}x$

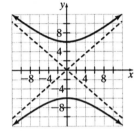

4. $y = \pm 4x$

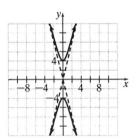

5. $\dfrac{x^2}{4} - \dfrac{y^2}{25} = 1;\ y = \pm\dfrac{5}{2}x$

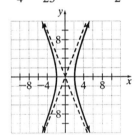

6. $\dfrac{x^2}{4} - \dfrac{y^2}{9} = 1;\ y = \pm\dfrac{3}{2}x$

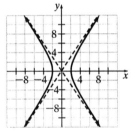

7. $\dfrac{x^2}{1} - \dfrac{y^2}{4} = 1;\ y = \pm 2x$

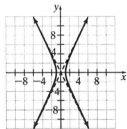

8. $\dfrac{y^2}{9} - \dfrac{x^2}{1} = 1;\ y = \pm 3x$

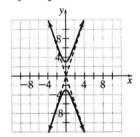

9.

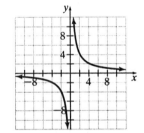

10.

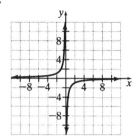

14.4 Nonlinear Systems of Equations and Their Applications

Summary

> A **system of nonlinear equations** is a system of
> equations in which at least one equation is not a linear
> equation (that is, one whose graph is not a straight line).
>
> One approach to solving a system of nonlinear equations
> is to graph its equation and determine the intersection of
> the graphs.

Example 1
Solve the system of equations graphically.

$$x^2 + y^2 = 4$$
$$x - y = 2$$

Solution
Graph each equation on the same axes and determine the intersection.

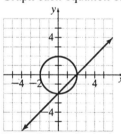

From the graph, the intersection is $(2, 0)$ and $(0, -2)$.

Check: $(2, 0)$: $\quad x^2 + y^2 = 4 \qquad\qquad x - y = 2$
$$2^2 + 0^2 = 4 \qquad\qquad 2 - 0 = 2$$
$$4 + 0 = 4 \qquad\qquad\quad 2 = 2 \text{ is true.}$$
$$4 = 4 \text{ is true.}$$

Check: $(0, -2)$: $\quad x^2 + y^2 = 4 \qquad\qquad x - y = 2$
$$0 + (-2)^2 = 4 \qquad\quad 0 - (-2) = 2$$
$$0 + 4 = 4 \qquad\qquad\quad 2 = 2 \text{ is true.}$$
$$4 = 4 \text{ is true.}$$

Summary

To solve a system of equations algebraically, we often solve one or more of the equations for one of the variables and then use substitute.

Example 2
Solve the system of equations algebraically using substitution.
$$2x^2 + y^2 = 9$$
$$2x - y = 3$$

Solution
Solve the linear equation for x or y. We select to solve for y (since its coefficient is -1).
$$2x - y = 3$$
$$-y = -2x + 3$$
$$y = 2x - 3$$

Substitute $y = 2x - 3$ for y in the equation $x^2 + y^2 = 9$ and solve for the remaining variable.
$$x^2 + y^2 = 9$$
$$x^2 + (2x - 3)^2 = 9$$
$$x^2 + 4x^2 - 12x + 9 = 9$$
$$5x^2 - 12x = 0$$
$$x(5x - 12) = 0$$
$$x = 0 \quad \text{or} \quad 5x - 12 = 0$$
$$5x = 12$$
$$x = \frac{12}{5}$$

Next, find the corresponding value of y by substituting each value of x in the equation solved for y.
If $x = 0$, and $y = 2x - 3$, then $y = 2(0) - 3 = -3$ so $(0, -3)$ is a solution.
If $x = \dfrac{12}{5}$, and $y = 2x - 3$, then $y = \dfrac{24}{5} - 3 = \dfrac{9}{5}$ so $\left(\dfrac{12}{5}, \dfrac{9}{5} \right)$ is a solution.

Summary

We can often solve systems of equations more easily using the addition method that was discussed in section 4.1. As with the substitution method, our objective is to obtain a single equation containing only one variable.

Example 3
Solve the system of equations using the addition method.
$$x^2 + y^2 = 26$$
$$x^2 - y^2 = 24$$

Solution

We can eliminate a variable by adding the two equations together.

$$x^2 + y^2 = 26$$
$$x^2 - y^2 = 24$$
$$2x^2 = 50$$
$$x^2 = 25$$
$$x = \pm\sqrt{25} = \pm 5$$

Finally, find the corresponding y values by substituting into the equation $x^2 + y^2 = 26$ and solving for y.

If $x = 5$, then $(5)^2 + y^2 = 26$
$$25 + y^2 = 26$$
$$y^2 = 1$$
$$y = \pm 1$$

Thus $(5, 1)$ and $(5, -1)$ are two ordered pair solutions.

If $x = -5$, then $(-5)^2 + y^2 = 26$
$$25 + y^2 = 26$$
$$y^2 = 1$$
$$y = \pm 1$$

So $(-5, 1)$ and $(-5, -1)$ are also solutions.

There are four solutions: $(5, 1), (5, -1), (-5, 1), (-5, -1)$.

Example 4

The area of a rectangle is 30 square feet. The perimeter is 26 feet. Find the length and width of the rectangle.

Solution

Let l = length of the rectangle and w = width of the rectangle.

Since the area of a rectangle is length times width, one equation is $lw = 30$.

The perimeter of a rectangle is the sum of twice the length and twice the width, thus a second equation is $2l + 2w = 26$.

Now, solve the system:
$$lw = 30$$
$$2l + 2w = 26$$

Solving the second equation for l we obtain

$$2l + 2w = 26$$
$$2l = 26 - 2w$$
$$l = 13 - w$$

Substituting that expression for l in the first equation, we obtain

$$lw = 30$$
$$(13 - w)(w) = 30$$
$$13w - w^2 = 30$$
$$0 = w^2 - 13w + 30$$
$$0 = (w - 10)(w - 3)$$

$w - 10 = 0$ or $w - 3 = 0$

    $w = 10$ or $w = 3$

If $w = 10$,   If $w = 3$
$l = 13 - w$ $l = 13 - w$
$l = 13 - 10 = 3$ $l = 13 - 3$
                     $l = 10$

So the dimensions of the rectangle are 3 feet by 10 feet.

Exercise Set 14.4

Solve the system of equations by the substitution method.

1. $y = x^2 + 2x$
    $y = x$

2. $y = x^2 - 4x + 4$
    $x + y = 2$

3. $x^2 + y^2 = 1$
    $x + 2y = 1$

4. $x - y = 2$
    $x^2 - y^2 = 16$

Solve the system of equations using the addition method.

5. $x^2 + y^2 = 4$
    $x^2 - y^2 = 4$

6. $2x^2 + 3y^2 = 6$
    $x^2 + 3y^2 = 3$

7. $5x^2 - 2y^2 = -13$
    $3x^2 + 4y^2 = 39$

8. $4x^2 + 4y^2 = 16$
    $2x^2 + 3y^2 = 5$

9. The area of a rectangle is 84 square feet. The perimeter is 38 feet. Find the dimensions of the rectangle.

Answers to Exercise Set 14.4

1. $(0, 0), (-1, -1)$

2. $(2, 0), (1, 1)$

3. $(1, 0), \left(-\dfrac{3}{5}, \dfrac{4}{5}\right)$

4. $(5, 3)$

5. $(2, 0), (-2, 0)$

6. $\left(\sqrt{3}, 0\right), \left(-\sqrt{3}, 0\right)$

7. $(1, 3), (1, -3), (-1, 3), (-1, -3)$

8. No real solution

9. 7 feet by 12 feet

Chapter 14 Practice Test

1. Write the equation of the circle with the center at $(-2, 4)$ and radius 5.

2. Write the equation in standard form. The, sketch the graph.
    $x^2 + y^2 - 4x + 8y - 5 = 0$

3. Sketch the graph of $9x^2 + 25y^2 = 225$.

4. Sketch the graph of $y = -3(x+2)^2 - 4$.

5. Sketch the graph of $x = y^2 + 4y + 3$.

6. Sketch the graph of $\dfrac{y^2}{9} - \dfrac{x^2}{81} = 1$.

7. Sketch the graph of $y = -\dfrac{6}{x}$.

8. Solve the system of equations.
$$4x^2 - y^2 = 7$$
$$y = 2x^2 - 3$$

9. Solve the system of equations.
$$6x^2 + y^2 = 9$$
$$3x^2 + 4y^2 = 36$$

10. The product of two numbers is 65. Their sum is 18. Find the numbers.

Answers to Chapter 14 Practice Test

1. $(x+2)^2 + (y-4)^2 = 25$ **2.** $(x-2)^2 + (y+4)^2 = 25$ **3.**

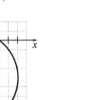
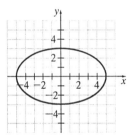

4.   **5.**   **6.**

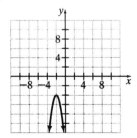
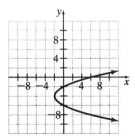
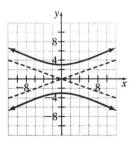

7.

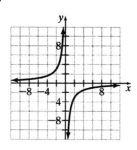

8. $\left(\sqrt{2}, 1\right), \left(-\sqrt{2}, 1\right)$

9. $(0, 3), (0, -3)$

10. 5 and 13

Chapter 15

15.1 Sequences and Series

Summary

Sequences

A **sequence (or progression)** of numbers is a list of numbers arranged in a specific order.

An **infinite sequence** is a function whose domain is the set of natural numbers.

A **finite sequence** is a function whose domain includes only the first n natural numbers.

Example 1

Write the first five terms of the sequence $a_n = n^2 - 2$.

Solution

$a_1 = 1^2 - 2 = -1$

$a_2 = 2^2 - 2 = 2$

$a_3 = 3^2 - 2 = 7$

$a_4 = 4^2 - 2 = 14$

$a_5 = 5^2 - 2 = 23$

Example 2

For the sequence $a_n = \dfrac{2n}{n+1}$, find

    a. the tenth term in the sequence.

    b. the fifteenth term in the sequence.

Solution

    a. For the tenth term, $n = 10$. So $a_{10} = \dfrac{2(10)}{10+1} = \dfrac{20}{11}$.

    b. For the fifteenth term, $n = 15$. Thus, $a_{15} = \dfrac{2(15)}{15+1} = \dfrac{30}{16} = \dfrac{15}{8}$.

Summary

> ### Series
>
> A **series** is the sum of the terms of a sequence. A **series** may be finite or infinite depending upon whether the sequence it is based on is finite or infinite.
>
> A **partial sum of a series** is the sum of a finite number of consecutive terms of the series, beginning with the first term.

Example 3
Given the sequence $a_n = 2n - 1$, find the first, third, and fifth partial sum.

Solution
For the first partial sum, we need $a_1 = 2(1) - 1 = 1$, $s_1 = 1$.
For the third partial sum, we need $a_1 = 2(1) - 1 = 1$, $a_2 = 2(2) - 1 = 3$ and $a_3 = 2(3) - 1 = 5$.
Now, $s_3 = 1 + 3 + 5 = 9$.
For the fifth partial sum, we need $a_1 = 1$, $a_2 = 3$, $a_3 = 5$, $a_4 = 7$, and $a_5 = 9$. $s_5 = 1 + 3 + 5 + 7 + 9 = 25$.

Summary

> A $\sum$ symbol is used to indicate which terms of a finite series are to be added.

Example 4
Write out the series and then find the sum of the series.

$$\sum_{k=1}^{4} (k^2 + 2)$$

Solution
Make a table:

k	$k^2 + 2$
1	$1^2 + 2 = 3$
2	$2^2 + 2 = 6$
3	$3^2 + 2 = 11$
4	$4^2 + 2 = 18$

$$\sum_{k=1}^{4} (k^2 + 2) = 3 + 6 + 11 + 18 = 38$$

Exercise Set 15.1

Write the first five terms of the sequence whose nth term is shown.

1. $a_n = 3n$

2. $a_n = n^2 + 5$

3. $a_n = \dfrac{2n}{n+1}$

Find the indicated term of the sequence whose nth term is shown.

4. $a_n = 4n - 3$; sixth term

5. $a_n = \dfrac{n^2}{n^2 + 1}$; fifth term

6. $a_n = (-1)n^3$; fourth term

Find the first, second, and fourth partial sums for the given sequence.

7. $a_n = 3n - 4$

8. $a_n = 2n^2 - 10$

9. $a_n = \left(\dfrac{1}{2}\right)^n$

Write out the series, and then find the sum of the series.

10. $\displaystyle\sum_{k=1}^{3} (2k + 3)$

11. $\displaystyle\sum_{k=1}^{5} \dfrac{k-1}{k}$

12. $\displaystyle\sum_{n=1}^{4} \dfrac{n(n+1)}{2}$

Answers to Exercise Set 15.1

1. $3, 6, 9, 12, 15$

2. $6, 9, 14, 21, 30$

3. $1, \dfrac{4}{3}, \dfrac{3}{2}, \dfrac{8}{5}, \dfrac{5}{3}$

4. $a_6 = 21$

5. $a_5 = \dfrac{25}{26}$

6. $a_4 = -64$

7. $s_1 = -1;\ s_2 = 1;\ s_4 = 14$

8. $s_1 = -8;\ s_2 = -10;\ s_4 = 20$

9. $s_1 = \dfrac{1}{2};\ s_2 = \dfrac{3}{4};\ s_4 = \dfrac{15}{16}$

10. $5 + 7 + 9 = 21$ **11.** $0 + \dfrac{1}{2} + \dfrac{2}{3} + \dfrac{3}{4} + \dfrac{4}{5} = \dfrac{163}{60}$ **12.** $1 + 3 + 6 + 10 = 20$

15.2 Arithmetic Sequences and Series

Summary

Arithmetic Sequence

A sequence in which each term after the first differs from the preceding term by a constant number is called an **arithmetic sequence** or **arithmetic progression.**

The constant amount by which each pair of successive terms differs is called the **common difference.**

nth term of an arithmetic sequence is $a_n = a_1 + (n-1)d$.

Example 1

    a. Write an expression for the general or nth term a_n, of the arithmetic sequence whose first term is 5 and whose common difference is –2.

    b. Find the tenth term of that sequence.

Solution

    a. Use $a_n = a_1 + (n-1)d$, with $a_1 = 5$ and $d = -2$.
$$a_n = 5 + (n-1)(-2)$$
$$a_n = 5 - 2n + 2$$
$$a_n = 7 - 2n$$

    b. Use $a_n = 7 - 2n$ with $n = 10$: $a_{10} = 7 - 2(10) = -13$.

Example 2
Find the number of terms in the arithmetic sequence 11, 16, 21, ..., 121.

Solution
The first term, a_1, is 11. The common difference, d, is 5. The nth term, a_n, is 121.
$$a_n = a_1 + (n-1)d$$
$$121 = 11 + (n-1)5$$
$$121 = 11 + 5n - 5$$
$$121 = 6 + 5n$$
$$115 = 5n$$
$$23 = n$$
The sequence has 23 terms.

Summary

> ### nth Partial Sum of an Arithmetic Sequence
>
> $$s_n = \frac{n(a_1 + a_n)}{2}$$

Example 3

Find the sum of the first 10 terms of the arithmetic sequence 4, 7, 10, ...

Solution

Use $s_n = \dfrac{n(a_1 + a_n)}{2}$. First, find a_n using $a_1 = 4$, $d = 3$ and $n = 10$.

$a_{10} = 4 + (10 - 1)(3) = 4 + 9(3) = 31$

Now, $s_{10} = \dfrac{10(4 + 31)}{2} = \dfrac{(10)(35)}{2} = 175$

Example 4

The first term of an arithmetic series is 1. The last term is 21. If $s_n = 121$, find the number of terms in the sequence and the common difference.

Solution

$a_1 = 1$, $a_n = 21$, $s_n = 121$

$s = \dfrac{n(a_1 + a_n)}{2}$

$121 = \dfrac{n(1 + 21)}{2} = \dfrac{n(22)}{2}$

$121 = 11n$

$n = 11$

There are 11 terms in the sequence.

Now,

$a_n = a_1 + (n - 1)d$

$21 = 1 + (11 - 1)d$

$21 = 1 + 10d$

$20 = 10d$

$2 = d$

The common difference is 2.

Example 5

If a person has a starting salary of \$1,000 per month and receives a \$20 raise each month, how much will have been earned in total at the end of 12 months?

Solution
Here, $a_1 = 1000$, $n = 12$, $d = 20$.
First, determine a_{12}.
$a_{12} = a_1 + (n-1)d$
$a_{12} = 1000 + (12-1)20$
$a_{12} = 1000 + 11(20)$
$a_{12} = 1220$

Now, use $s_n = \dfrac{n(a_1 + a_n)}{2} = \dfrac{12(1000 + 1220)}{2} = \dfrac{12(2220)}{2} = 13,320.$
A total of $13,320 will have been earned.

Exercise Set 15.2

Write the first five terms and an expression for the nth term of the arithmetic sequence with the given first term and common difference.

1. $a_1 = 1$, $d = -2$ 　　　　**2.** $a_1 = -9$, $d = 3$ 　　　　**3.** $a_1 = 3$, $d = \dfrac{3}{4}$

Find the desired quantity of the arithmetic sequence.

4. $a_1 = 4$, $d = 3$, find a_{25}. 　　**5.** $a_1 = 5$, $d = 4$, $a_n = 77$, find n. 　**6.** $a_1 = 2$, $a_{24} = 48$, find d.

Find the sum, s_n, and the common difference, d.

7. $a_1 = 4$, $a_n = 28$, $n = 7$ 　　　　**8.** $a_1 = 3$, $a_n = -84$, $n = 30$

9. A seating section in a theater-in-the-round has 20 seats in the front row, 22 in the second row, 24 in the third row and so on for 25 rows. How many seats are there in the last row? How many seats are there in the section?

Answers to Exercise Set 15.2

1. $1, -1, -3, -5, -7$; $a_n = 3 - 2n$

2. $-9, -6, -3, 0, 3$; $a_n = 3n - 12$

3. $3, \dfrac{15}{4}, \dfrac{9}{2}, \dfrac{21}{4}, 6$; $a_n = \dfrac{9}{4} + \dfrac{3}{4}n$

4. 76 　　　　　　　　**5.** 19 　　　　　　　　**6.** 2

7. $d = 4$; $s_7 = 112$ 　　　　**8.** $d = -3$; $s_{30} = -1215$

9. 68 seats in the last row; 1100 seats in the section.

15.3 Geometric Sequences and Series

Summary

> **Geometric Sequence**
>
> A **geometric sequence** (or **geometric progression**) is a sequence in which each term after the first is a multiple of the preceding term.
>
> The common multiple is called the **common ratio, r.**

Example 1
Determine the first four terms of the geometric sequence if $a_1 = 1$ and $r = 2$.

Solution
$a_1 = 1,\ a_2 = 1 \cdot 2 = 2,\ a_3 = 2 \cdot 2 = 4,\ a_4 = 4 \cdot 2 = 8,\ a_5 = 8 \cdot 2 = 16$
The first five terms are 1, 2, 4, 8, and 16.

Summary

> **nth Term of a Geometric Sequence**
>
> $$a_n = a_1 \cdot r^{n-1}$$

Example 2

    a. Write an expression for the general (or nth) term, a_n, of the geometric sequence with $a_1 = 2$ and $r = -3$.

    b. Find the eight term of the sequence.

Solution

    a. $a_1 = 2,\ r = -3$ so
      $a_n = a_1 \cdot r^{n-1}$
      $a_n = 2 \cdot (-3)^{n-1}$

    b. $a_n = 2 \cdot (-3)^{n-1},\ a_8 = 2(-3)^{(8-1)} = 2(-3)^7 = -4374$

Example 3
Find r and a_1 for the geometric sequence with $a_2 = 6$ and $a_6 = 486$.

Solution

Consider the sequence ___, 6, ___, ___, ___, 486.

Let $a_1 = 6$. Therefore, $a_5 = 486$. So, $a_n = a_1 \cdot r^{n-1}$

$$486 = 6 \cdot r^{5-1}$$
$$486 = 6 \cdot r^4$$
$$81 = r^4$$
$$3 = r$$

To find a_1, use $a_n = a_1 \cdot r^{n-1}$

$$a_2 = a_1 3^{2-1}$$
$$6 = a_1 3^1$$
$$2 = a_1$$

Summary

---

nth Partial Sum of a Geometric Series

$$s_n = \frac{a_1(1-r^n)}{1-r}, \; r \neq 1$$

Example 4

Find the sixth partial sum of a geometric sequence whose first term is 27 and whose common ratio is $\frac{1}{3}$.

Solution

$$s_n = \frac{a_1(1-r^n)}{1-r}$$

$$s_6 = \frac{27\left(1-\left(\frac{1}{3}\right)^6\right)}{1-\frac{1}{3}} = \frac{27\left(1-\frac{1}{729}\right)}{1-\frac{1}{3}}$$

$$s_6 = \frac{27\left(\frac{728}{729}\right)}{\frac{2}{3}} = \frac{364}{9}$$

Example 5
Given $s_n = 363$, $a_1 = 3$, $r = 3$, find n.

Solution

$$s_n = \frac{a_1(1 - r^n)}{1 - r}$$

$$363 = \frac{3(1 - 3^n)}{1 - 3}$$

$$363 = \frac{3(1 - 3^n)}{-2}$$

$$-242 = 1 - 3^n$$

$$-243 = -3^n$$

$$243 = 3^n$$

$$5 = n$$

Example 6
A city has a population of 100,000. If the population grows at a rate of 5% per year, find

    a. the population in 10 years.

    b. the number of years for the population to double.

Solution

    a. Here, $a_1 = 100,000$, $r = 1.05$, $n = 10$.

$$a_n = a_1 \cdot r^{n-1}$$

$$a_{10} = 100,000 \cdot (1.05)^{10-1} \approx 155,133$$

    b. Here, $a_1 = 100,000$, $a_n = 200,000$, $r = 1.05$ and n is unknown.

Use the formula $a_n = a_1 \cdot r^{n-1}$.

$$200,000 = 100,000 \cdot (1.05)^{n-1}$$

$$2 = (1.05)^{n-1}$$

$$\ln 2 = (n - 1) \cdot \ln(1.05)$$

$$\frac{\ln 2}{\ln 1.05} = n - 1$$

$$\frac{\ln 2}{\ln 1.05} + 1 = n$$

$$15.2 = n$$

The population will double in a little more than 15 years.

Summary

Example 7

Find the sum of the terms of the sequence $4, \ \dfrac{2}{3}, \ \dfrac{1}{9}, \ \ldots$

Solution

$a_1 = 4, \ r = \dfrac{1}{6}$, since $4 \cdot \dfrac{1}{6} = \dfrac{2}{3}$.

Since $|r| < 1$, the formula for the sum of an infinite geometric series can be used:

$$s_\infty = \frac{a_1}{1-r} = \frac{4}{1-\frac{1}{6}} = \frac{4}{\frac{5}{6}} = 4 \cdot \frac{6}{5} = \frac{24}{5} = 4.8$$

Example 8

Find the sum of the infinite series below.

$$3 + \frac{3}{5} + \frac{3}{25} + \cdots$$

Solution

$a_1 = 1, \ r = \dfrac{1}{5}$, since $3 \cdot \dfrac{1}{5} = \dfrac{3}{5}$. Since $|r| < 1$, the formula for the sum of an infinite geometric series can be used

once again:

$$s_\infty = \frac{a_1}{1-r} = \frac{3}{1-\frac{1}{5}} = \frac{3}{\frac{4}{5}} = 3 \cdot \frac{5}{4} = \frac{15}{4} = 3\frac{3}{4} = 3.75$$

Example 9

Write the rational number $0.121212\ldots$ as a ratio of two integers using an infinite geometric series.

Solution

The decimal number $0.121212\ldots$ can be written as a series: $0.12 + 0.0012 + 0.000012 + \cdots$

Here, $a_1 = 0.12$ and $r = 0.01$ and the formula for the sum of an infinite geometric series can again be used:

$$s_\infty = \frac{a_1}{1-r} = \frac{0.12}{1-0.01} = \frac{0.12}{0.99} = \frac{12}{99} = \frac{4}{33}$$

Exercise Set 15.3

Determine the first five terms of the geometric sequence.

1. $a_1 = 4, \ r = 2$ **2.** $a_1 = 81, \ r = -\dfrac{1}{3}$

Find the indicated term of the geometric sequence.

3. $a_1 = 5$, $r = \dfrac{1}{2}$, find a_5.   **4.** $a_1 = 1$, $r = 5$, find a_8.

Find the sum.

5. $a_1 = -2$, $r = 3$, find s_6.   **6.** $a_1 = 12$, $r = -\dfrac{1}{2}$, find s_5.

7. In a geometric sequence $a_3 = -12$ and $a_6 = 96$. Find r and a_1.

8. In a geometric sequence, $a_2 = 64$, $a_8 = 1$, find r and a_1.

9. The population of a city is 500,000 and grows at the rate of 4% per year.

   a. What will the population be in 10 years?

   b. In how many years will the population double?

Find the sums.

10. $7 + \dfrac{7}{3} + \dfrac{7}{9} + \cdots$ **11.** $\dfrac{5}{3} + \dfrac{5}{12} + \dfrac{5}{48} + \cdots$ **12.** $243 + 162 + 108 + \cdots$

13. $16 + 12 + 9 + \cdots$ **14.** $10 - 5 + \dfrac{5}{2} - \dfrac{5}{4} + \cdots$

Write the decimal number as a ratio of two integers.

15. $0.232323\ldots$ **16.** $0.123123123\ldots$

Answers to Exercise Set 15.3

   1. 4, 8, 16, 32, 64   **2.** 81, –27, 9, –3, 1   **3.** $\dfrac{5}{16}$

   4. 78,125   **5.** –728   **6.** $\dfrac{33}{4}$

   7. $a_1 = -3$, $r = -2$ **8.** $a_1 = 128$, $r = \dfrac{1}{2}$

   9. a. 711,656   **b.** 18.7 years

   10. $\dfrac{21}{2}$ **11.** $\dfrac{20}{9}$ **12.** 729

13. 64

14. $\dfrac{20}{3}$

15. $\dfrac{23}{99}$

16. $\dfrac{123}{999}$

15.4 The Binomial Theorem

Summary

---
n Factorial

$n! = n(n-1)(n-2)(n-3) \ldots (1)$
for any positive integer n

$0! = 1$

Example 1
Find

   a. 4!

       b. 7!

Solution

   a. $4! = 4 \cdot 3 \cdot 2 \cdot 1 = 24$

   b. $7! = 7 \cdot 6 \cdot 5 \cdot 4 \cdot 3 \cdot 2 \cdot 1 = 5040$

Summary

Binomial Coefficients

For n and r nonnegative integers, $n > r$

$$\binom{n}{r} = \frac{n!}{r! \cdot (n-r)!}$$

Example 2
Evaluate

   a. $\dbinom{7}{3}$

       b. $\dbinom{8}{4}$

   c. $\dbinom{6}{0}$

       d. $\dbinom{6}{6}$

Solution

a. $\dbinom{7}{3} = \dfrac{7!}{3!(7-3)!} = \dfrac{7 \cdot 6 \cdot 5 \cdot 4 \cdot 3 \cdot 2 \cdot 1}{3 \cdot 2 \cdot 1 \cdot 4 \cdot 3 \cdot 2 \cdot 1} = 35$

b. $\dbinom{8}{4} = \dfrac{8!}{4!(8-4)!} = \dfrac{8 \cdot 7 \cdot 6 \cdot 5 \cdot 4 \cdot 3 \cdot 2 \cdot 1}{4 \cdot 3 \cdot 2 \cdot 1 \cdot 4 \cdot 3 \cdot 2 \cdot 1} = 70$

c. $\dbinom{6}{0} = \dfrac{6!}{0!(6-0)!} = \dfrac{6!}{6!} = 1$

d. $\dbinom{6}{6} = \dfrac{6!}{6!(6-6)!} = \dfrac{6!}{6! \cdot 0!} = 1$

Summary

Binomial Theorem

For any positive integer n,

$$(a+b)^n = \binom{n}{0}a^n \cdot b^0 + \binom{n}{1}a^{n-1} \cdot b^1 + \binom{n}{2}a^{n-2} \cdot b^2 + \binom{n}{3}a^{n-3} \cdot b^3 + \cdots + \binom{n}{n}a^0 \cdot b^n$$

Example 3

Expand $(3x + y)^5$ using the binomial formula.

Solution

Use $(a+b)^n = \binom{n}{0}a^n \cdot b^0 + \binom{n}{1}a^{n-1} \cdot b^1 + \binom{n}{2}a^{n-2} \cdot b^2 + \binom{n}{3}a^{n-3} \cdot b^3 + \cdots + \binom{n}{n}a^0 \cdot b^n$

$a = 3x$, $b = y$ and $n = 5$:

$\binom{5}{0}(3x)^5 y^0 + \binom{5}{1}(3x)^4 y^1 + \binom{5}{2}(3x)^3 y^2 + \binom{5}{3}(3x)^2 y^3 + \binom{5}{4}(3x)^1 y^4 + \binom{5}{5}(3x)^0 y^5$

$= 3^5 x^5 + 5(3^4)x^4 y + 10(3^3)x^3 y^2 + 10(3^2)x^2 y^3 + 5(3x)y^4 + 1 \cdot y^5$

$= 243x^5 + 405x^4 y + 270x^3 y^2 + 90x^2 y^3 + 15xy^4 + y^5$

Exercise Set 15.4

Evaluate the following.

1. $3!$

2. $5!$

3. $\dbinom{4}{0}$

4. $\dbinom{4}{1}$

5. $\dbinom{7}{2}$

6. $\dbinom{8}{3}$

Use the binomial formula to expand each of the following.

7. $(r+s)^3$ **8.** $(a-b)^4$ **9.** $(3x-y)^5$

10. $(2x+3y)^4$

Answers to Exercise Set 15.4

1. 6   **2.** 120   **3.** 1

4. 4   **5.** 21   **6.** 56

7. $r^3 + 3r^2s + 3rs^2 + s^2$ **8.** $a^4 - 4a^3b + 6a^2b^2 - 4ab^3 + b^4$

9. $243x^5 - 405x^4y + 270x^3y^2 - 90x^2y^3 + 15xy^4 - y^5$

10. $16x^4 + 96x^3y + 216x^2y^2 + 216xy^3 + 81y^4$

Chapter 15 Practice Test

1. Write the first five terms of the sequence with $a_n = \dfrac{2n}{n+1}$.

2. Find the first and third partial sum of $a_n = \dfrac{n^2}{n+4}$.

3. Write out the series and find the sum of the series $\displaystyle\sum_{n=1}^{5}(3n-2)$.

4. Write the general term for the arithmetic sequence.
4, 11, 18, ...

5. Write the general term for the geometric sequence.
12, 6, 3, ...

6. Write the first four terms of the sequence where $a_1 = 10$, $d = -4$.

7. Write the first four terms of the sequence where $a_1 = 8$ and $r = \dfrac{3}{4}$.

8. Find a_8 when $a_1 = -3$ and $d = -4$.

9. Find s_{10} if $a_1 = 8$ and $d = 3$.

10. Find the number of terms in the arithmetic sequence.
-3, 1, 5, 9, ..., 45

11. Find a_6 if $a_1 = 8$ and $r = -\dfrac{1}{2}$.

12. Find s_4 if $a_1 = 10$ and $r = \dfrac{1}{5}$.

13. Find the common ratio and write a general term for the sequence.
$8, -4, 2, -1$

14. Find the sum of the infinite geometric series.
$10 + 2 + \dfrac{2}{5} \cdots$

15. Use the binomial theorem to expand $(2x - y)^5$.

Answers to Chapter 15 Practice Test

1. $1, \dfrac{4}{3}, \dfrac{3}{2}, \dfrac{8}{5}, \dfrac{5}{3}$

2. $s_1 = \dfrac{1}{5}, s_3 = \dfrac{226}{105}$

3. $1 + 4 + 7 + 10 + 13 = 35$

4. $a_n = 7n - 3$

5. $a_n = 12\left(\dfrac{1}{2}\right)^{n-1}$

6. $10, 6, 2, -2$

7. $8, 6, \dfrac{9}{2}, \dfrac{27}{8}$

8. -31

9. 215

10. 13

11. $-\dfrac{1}{4}$

12. $\dfrac{312}{25}$

13. $r = -\dfrac{1}{2}; a_n = 8\left(-\dfrac{1}{2}\right)^{n-1}$

14. $\dfrac{25}{2}$

15. $32x^5 - 80x^4 y + 80x^3 y^2 - 40x^2 y^3 + 10xy^4 - y^5$